THE SCIENCE HANDBOOK

INTRODUCING LAWS, THEORIES, AND PRINCIPLES OF GEOGRAPHY, PHYSICS, CHEMISTRY, BIOLOGY AND GEOLOGY

MATTHEW S. DEAL

CONTENTS

Preface   VII

Copyrights   VIII

Part 1

Geography and Related Studies   3

1. First Law of Geography   5

2. Central Place Theory   9

3. Theory of Environmental Determinism   13

4. Theory of Possibilism   17

5. Law of Retail Gravitation   21

6. Principle of Least Effort   25

7. Theory of Human Ecology   29

8. Theory of Spatial Diffusion   33

Part 2

Physics   41

9. Law of Universal Gravitation   43

10. Theory of Relativity   47

11. Quantum Mechanics   51

12. Law of Large Numbers   55

13. The Central Limit Theorem   59

14. The Principle of Least Action   65

15. The Big Bang Theory   69

16. Newton's Laws   73

17. First Law of Thermodynamics   85

18. Laws of Cartoon Physics – not meant to be taken seriously   89

Part 3

Chemistry   97

19. Law of Conservation of Mass   99

20. Law of Multiple Proportions   103

21. Law of Definite Proportions   107

22. Theories of Valence   111

23. Theories of Molecular Orbitals   115

24. Theories of Chemical Bonding   119

Part 4

Biology   127

25. Cell Theory   129

26. Theory of Evolution   133

27. Theory of Genetics   137

28. Theory of Ecology   141

29. Theory of Homeostasis   147

30. Law of the Minimum   151

31. Theory of Growth and Development   155

Part 5

Geology   163

32. Theory of Plate Tectonics   165

33. Theory of the Evolution of the Earth   169

34. Theory of the Formation of Rocks   175

35. Theory of the Formation of Minerals   181

About the Author   185

Afterword   186

To my daughter,

This book, an introduction to a few concepts of science, is a modest attempt to capture the wonders of the universe. Yet, the greatest wonder I've ever known is watching you grow.

May the pages of this book inspire in you an unbounded curiosity, just as you have inspired me. And as you journey through life, may you always be reminded of the infinite possibilities that lie ahead, fueled by passion, and guided by knowledge.

PREFACE

This handbook presents a diverse array of scientific concepts, with the intent that one or several will spark an interest in deeper exploration. My first attempt for this book was to compile all the "First Laws" of science as a gift for my daughter, hoping she'd discover a topic that piqued her interest. However, focusing solely on the "first laws" omitted too many fascinating and crucial concepts that would be ideal in an introduction handbook such as this. Consequently, the book evolved to feature various theories and principles of science, aiming to attract folks with a term, concept, or idea that drives their curiosity further. Thus, this book is not intended as a reference source, but rather as a catalyst for exploration and discovery.

PART 1

GEOGRAPHY

GEOGRAPHY AND RELATED STUDIES

*G*eography is the study of Earth's landscapes, environments, and most importantly, the relationships between things, like people and their environments. It's a broad discipline that bridges the physical and social sciences. Physical geography investigates the natural world, including landforms, climates, ecosystems, and the processes that shape them. Human geography, on the other hand, examines the impact of human societies on the planet and explores cultural, economic, political, and demographic phenomena. The discipline also encompasses the study of spatial aspects of human existence - how places and spaces shape our experiences and influence various socioeconomic and political phenomena.

Geography's importance cannot be overstated. It is critical in understanding some of the most pressing challenges of the 21st century, including climate change, environmental degradation, urbanization, and global inequality. By understanding the geographical factors influencing these issues, we can develop effective strategies to alleviate their impacts. Furthermore, geography helps us appreciate the diversity of cultures, political systems, economies, and environments across the world, fostering global awareness and empathy. It also contributes to a range of practical fields, including urban planning, transportation, tourism, international development, and environmental conservation. In essence, *studying geography equips us with knowledge and skills that are valu-able for navigating our interconnected world and working towards a sustainable future.*

And so, we start with some of the theories and laws of geography that hold immense importance and how location influences the diverse aspects of our lives.

1

FIRST LAW OF GEOGRAPHY

Tobler's Law, also known as Tobler's First Law of Geography, is a fundamental principle in spatial analysis. This law highlights the significance of proximity and spatial relationships in understanding geographic patterns.

Definition of Tobler's Law: Tobler's Law can be summarized as follows: "Everything is related to everything else, but near things are more related than distant things." This principle suggests that the interaction and relationship between things in space are influenced by their proximity. Objects that are geographically closer to each other are more likely to have stronger and more frequent interactions compared to those that are farther apart.

Credit to Waldo Tobler: Tobler's Law is named after Waldo Tobler, an esteemed geographer and cartographer who made significant contributions to the field of spatial analysis. Tobler formulated this law in the early 1970s based on his research and understanding of spatial relationships and spatial autocorrelation. His work has had a profound impact on the field of geography and has shaped the way researchers analyze and interpret spatial data.

Example of Tobler's Law in Real Life: An example of Tobler's Law in real life is the study of commuting patterns. When examining the movement of individuals between their residences and workplaces, it is evident that people tend to have a higher probability of commuting to destinations that are closer to their homes. This aligns with Tobler's Law, as the proximity of the residence and workplace facilitates more frequent interactions and easier commuting. The law helps explain the phenomenon of commuters favoring jobs located near their homes and the existence of commuting patterns that decrease with increasing distance.

Importance of Tobler's Law: Tobler's Law holds significant importance in various aspects of geography and spatial analysis:

1. Spatial Interaction: Tobler's Law helps us understand the patterns and intensity of spatial interactions. By considering the proximity of locations, researchers can better model and predict the flow of people, goods, and information between places. This knowledge is crucial in urban planning, transportation studies, and the analysis of trade networks.

2. Diffusion of Phenomena: The law contributes to our understanding of the diffusion and spread of phenomena. When studying the adoption of innovations, the spread of diseases, or the dissemination of cultural practices, Tobler's Law reminds us to consider the role of geographic proximity in influencing the speed and extent of diffusion.

3. Spatial Data Analysis: Tobler's Law underscores the importance of accounting for spatial autocorrelation when analyzing and interpreting spatial data. Spatial autocorrelation

refers to the tendency of neighboring observations to exhibit similar values. By recognizing the influence of Tobler's Law, researchers can employ appropriate statistical techniques that account for spatial dependencies, leading to more accurate and meaningful analysis.

4. Urban Planning and Policy Making: The law has implications for urban planning and policy decisions. Understanding the impact of proximity on various urban relationships, such as access to services, socio-economic disparities, and transportation infrastructure, can inform more effective planning and resource allocation.

Conclusion: Tobler's Law highlights the importance of proximity and spatial relationships in understanding geographic phenomena. The law asserts that near things are more related than distant things, emphasizing the influence of spatial context on interactions and patterns in space. It has profound implications for spatial analysis, spatial data modeling, diffusion studies, and urban planning. By considering Tobler's Law, researchers gain valuable insights into the relationships between phenomena, helping them make informed decisions and predictions about spatial processes. The significance of Tobler's Law lies in its ability to guide our understanding of spatial interactions, spatial data analysis, and the complexity of spatial patterns in our interconnected world.

2

Central Place Theory

Central Place Theory is a fundamental concept in human geography that seeks to explain the spatial distribution and hierarchy of human settlements. This theory provides insights into the organization of urban centers based on their size, function, and spatial relationships.

Definition of Central Place Theory: Central Place Theory aims to explain the spatial arrangement and economic relationships of human settlements. According to this theory, settlements serve as central places that provide goods and services to the surrounding population. The theory proposes a hierarchical structure of urban centers, with larger and more influential central places serving larger market areas and providing higher-order goods and services. The theory also emphasizes the importance of distance and market thresholds in shaping the pattern of settlement distribution.

Imagine you have a bunch of towns on a big map, and each town offers different goods and services to the people living there. Some towns are small and offer basic things like groceries, while others are bigger and provide more specialized items like electronics or clothes.

The theory says that these towns are like "central places" that attract people from surrounding areas. The bigger and more important a town is, the farther people are willing to travel to get there. Smaller towns are closer together because they offer everyday things that people need more frequently.

According to Central Place Theory, there is a hierarchy of towns based on the goods and services they provide. Large cities are at the top of the hierarchy because they offer a wide range of products and services, attracting people from far away. Smaller towns are below them, providing more basic things to the local population.

This theory helps urban planners and geographers understand how towns and cities function in relation to each other and how they meet the needs of the people living in the area. It also helps us see patterns in how towns are spaced out and why some places are more important for trade and commerce than others.

Credit to Walter Christaller: Central Place Theory is credited to Walter Christaller, a German geographer who developed the theory in the 1930s. Christaller sought to create a model that would explain the regularity and hierarchy of settlements based on economic principles. His work laid the foundation for understanding the functional relationships between urban centers and their surrounding regions.

Example of Central Place Theory in Real Life: A practical example of Central Place Theory is observed in the retail sector. In a given area, larger urban centers, such as cities, act as central places that offer a wide range of goods and services, including high-order items like luxury goods and specialized professional services. Smaller towns and villages, on the other hand, serve as lower-order central places that provide essential goods and services for the local population. The theory predicts that the larger urban centers will have a wider sphere of influence, serving a larger population and offering a greater variety of goods and services, while smaller towns will have a more limited influence and cater to the immediate surrounding areas.

Importance of Central Place Theory: Central Place Theory holds significant importance in various aspects of urban geography and urban planning:

1. Urban Hierarchy and Market Analysis: The theory helps in understanding the hierarchical structure of urban centers, the organization of goods and services, and the functional relationships between settlements. It aids in market analysis, allowing businesses and policymakers to assess the potential market size and the viability of various goods and services in specific locations.

2. Transportation Planning and Accessibility: Central Place Theory informs transportation planning and infrastructure development. By understanding the spatial distribution of central places, planners can identify transportation needs, determine optimal locations for transportation nodes, and improve accessibility to ensure efficient movement of people and goods within the urban hierarchy.

3. Regional Planning and Resource Allocation: The theory aids in regional planning by providing insights into the provision of goods and services across different areas. It helps policymakers in resource allocation, ensuring that the needs of the population are met in an equitable manner and that

rural areas have access to essential services.

4. Retail and Service Sector Analysis: Central Place Theory has practical applications in the retail and service sectors. It helps retailers and service providers identify optimal locations for their establishments, considering factors such as market size, competition, and catchment areas. The theory assists in understanding the spatial organization of retail and service facilities within urban and rural areas.

Conclusion: Central Place Theory provides a valuable framework for understanding the organization, hierarchy, and economic relationships of human settlements. The theory explains how urban centers function as central places, serving the needs of surrounding populations. By recognizing the spatial distribution of central places and their varying functions, policymakers, planners, and businesses can make informed decisions regarding transportation planning, resource allocation, market analysis, and retail/service sector development. Central Place Theory remains a cornerstone in urban geography, offering insights into the spatial organization and dynamics of human settlements in both urban and rural contexts.

3

***

Theory of Environmental Determinism

The Theory of Environmental Determinism, also known as Climatic Determinism, was a concept that sought to explain the influence of the natural environment on human societies. Developed during the 19th and early 20th centuries, this theory posited that the physical environment, including climate, topography, and resources, determined the cultural, social, and economic development of civilizations.

Definition of the Theory of Environmental Determinism: The Theory of Environmental Determinism asserts that the natural environment, particularly climate and geography, is the primary factor shaping the characteristics and progress of human societies. According to this theory, environmental conditions dictate the development of cultural practices, technology, economic systems, and societal structures. It suggests that different environments produce distinct civilizations and societies due to their inherent natural advantages and limitations.

In other words it suggests our natural surroundings, such as climate, geography, and resources, strongly influence and even control human behavior and development. In simple terms, it means that where we live and the environment we are in shape who we are and how we live.

According to environmental determinism, factors like the weather, the availability of water and fertile land, and the types of resources in an area can determine things like the type of houses people build, the food they eat, and the jobs they do. For example, in hot and arid regions, people might build houses with thick walls to stay cool, and they might rely more on farming that suits the dry climate. On the other hand, in colder regions, people might build houses with fireplaces and focus on activities like hunting and fishing to adapt to the environment.

However, it is essential to note that environmental determinism has been widely criticized because it oversimplifies the complexity of human societies and their ability to adapt and change. Human behavior is also influenced by culture, technology, and social factors, which can sometimes override the limitations of the environment. Therefore, while the environment can be an important factor, it is not the only thing that determines how people live and behave.

Credit to Friedrich Ratzel, Ellen Churchill Semple, and Historical Context: The Theory of Environmental Determinism can be attributed to several key contributors, including Friedrich Ratzel, a German geographer, and Ellen Churchill Semple, an American geographer. Ratzel's work on anthropogeography and Semple's writings on the influence of the environment on civilizations helped popularize this theory during the late 19th and early 20th centuries. Their works were influenced by prevailing scientific and social perspectives of the time, including Darwin's theory of evolution and the emphasis on racial and cultural superiority.

Example of the Theory of Environmental Determinism in Real Life: An example of the Theory of Environmental Determinism can be observed in the historical analysis of agricultural practices. The theory suggests that societies in arid regions developed irrigation systems and focused on agriculture as a means of survival due to the scarcity of water and limited fertile land. In contrast, societies in regions with abundant rainfall and fertile soil were more likely to rely on agriculture as their primary economic activity. This example demonstrates how environmental factors, such as climate and natural resources, influenced the development of specific cultural practices and economic systems.

Importance of the Theory of Environmental Determinism: The Theory of Environmental Determinism, despite being heavily criticized and largely discredited today, played a significant role in the development of geography and our understanding of the relationship between humans and their environment:

1. Historical Context and Intellectual Development: The theory emerged during a time when scientific exploration and geographic discoveries were taking place. It contributed to the development of geography as a discipline and sparked discussions about the interconnectedness of the natural environment and human societies.

2. Early Geographic Thought: The Theory of Environmental Determinism provided early geographers with a framework to study the impact of physical environment on human activities. It prompted researchers to examine patterns of settlement, cultural practices, and economic systems in relation to environmental conditions.

3. Evolution of Geographical Thought: Critiques and challenges against the deterministic perspective led to the evolution of geographical thought. The theory's shortcomings encouraged geographers to explore other factors, such as human agency, cultural diversity, and historical processes, in understanding the complex relationship between environment and society.

4. Contemporary Critique and Transformation: While the theory has largely been discredited, it played a crucial role in sparking critical debates and moving the field of geography towards more nuanced and holistic perspectives. It led to the emergence of alternative theories, such as possibilism, which argue that human societies possess the capacity to adapt to and modify their environments.

Conclusion: The Theory of Environmental Determinism posited that the natural environment was the primary determinant of human civilization and cultural development. Although it faced significant critique and has largely been discredited, this theory played an essential role in shaping early geographic thought and our understanding of the interplay between environment and society. The theory's historical significance lies in the intellectual development of geography and the subsequent evolution of more nuanced and multidimensional theories that consider the complex interactions between humans and their environment.

4

THEORY OF POSSIBILISM

The Theory of Possibilism emerged as a response to the deterministic perspectives of Environmental Determinism. Unlike Environmental Determinism, which proposed that the natural environment determines the development of human societies, Possibilism asserts that humans have the ability to adapt to and modify their environment through cultural practices and technological advancements.

Definition of the Theory of Possibilism: The Theory of Possibilism posits that human societies possess the agency to shape their environment and overcome the limitations imposed by physical conditions. It recognizes that while the natural environment provides a range of possibilities, human culture, innovation, and technological advancements allow societies to adapt to diverse environments and alter their surroundings to better suit their needs. Possibilism emphasizes the active role of humans in modifying and transforming the environment through their actions.

It suggests that while the natural environment does influence human life, people have the ability to adapt, innovate, and make choices that go beyond the limitations imposed by the environment and in simple terms, possibilism means that humans have the power to make possibilities or choices based on their needs, desires, and available technology, even in challenging environments. It acknowledges that people can find creative solutions to overcome environmental constraints.

For example, in a cold and mountainous region, environmental determinism might suggest that people would only engage in limited activities due to the harsh climate and rugged terrain. However, possibilism recognizes that humans can build warm and sturdy homes, invent suitable clothing, and use modern transportation to connect with other places and access resources, expanding their possibilities for living in such an environment.

In essence, the theory of possibilism highlights human ingenuity and adaptability, emphasizing that our actions and choices play a vital role in shaping how we live and interact with the environment, rather than being entirely determined by it.

Credit to Vidal de la Blache and Historical Context: The Theory of Possibilism is credited to French geographer Paul Vidal de la Blache, who proposed this concept in the late 19th and early 20th centuries. Vidal de la Blache rejected the deterministic views of Environmental Determinism and advocated for a more nuanced understanding of the human-environment relationship. His ideas became influential in the field of geography, challenging the deterministic perspective prevalent at the time.

Example of the Theory of Possibilism in Real Life: An example of the Theory of Possibilism can be observed in the development of agricultural practices. In areas with limited fertile soil, communities have adapted by implementing innovative techniques such as terracing, crop rotation, and irrigation systems to maximize agricultural productivity. By utilizing their cultural knowledge and technological advancements, societies can overcome environmental limitations and create possibilities for sustainable agricultural practices even in challenging landscapes. This example demonstrates how human agency and innovation play a vital role in shaping the environment to suit human needs.

Importance of the Theory of Possibilism: The Theory of Possibilism holds significant importance in several areas of study and understanding the human-environment relationship:

1. Human-Centered Perspective: Possibilism places humans at the center of the environmental narrative, emphasizing their ability to interact with and modify the environment. It recognizes the diverse ways in which human societies have adapted and overcome environmental constraints through cultural practices, technology, and innovation.

2. Cultural Diversity and Adaptation: Possibilism acknowledges the cultural diversity and variability in human responses to the environment. It highlights how different societies develop unique ways of adapting to their specific environmental conditions, showcasing the rich tapestry of human creativity and ingenuity.

3. Sustainable Development: The theory emphasizes the potential for sustainable development by focusing on human agency and innovative practices. By understanding the possibilities for environmental adaptation and modification, societies can strive for sustainable solutions that balance human needs with the preservation and conservation of natural

resources.

4. Environmental Management and Planning: Possibilism offers insights into environmental management and planning. Recognizing human agency in shaping the environment prompts policymakers and planners to consider the social, cultural, and technological factors that influence the interaction between humans and their surroundings. This perspective guides decision-making processes that promote sustainable land use, resource management, and urban development.

Conclusion: The Theory of Possibilism presents an alternative perspective to the deterministic views of Environmental Determinism. By emphasizing human agency and innovation, Possibilism recognizes the capacity of human societies to shape and adapt to their environment. This theory highlights the importance of cultural practices, technology, and human creativity in overcoming environmental limitations. Understanding the Theory of Possibilism allows us to appreciate the diverse ways in which human societies interact with and modify their surroundings, promoting sustainable development and harmonious human-environment relationships.

5

LAW OF RETAIL GRAVITATION

The Law of Retail Gravitation is a principle that explains consumer behavior and retail patterns based on the distance and attractiveness of retail centers. It suggests that consumers are drawn to retail centers based on their distance and the attractiveness of the center, including factors like variety, price, and convenience.

Definition of the Law of Retail Gravitation: The Law of Retail Gravitation posits that consumers are inclined to patronize retail centers based on their distance and attractiveness. According to this principle, the size and attractiveness of retail centers, combined with the distance consumers need to travel, influence their shopping preferences. Consumers are more likely to be drawn to larger, more attractive retail centers that offer a variety of goods, competitive prices, and convenient amenities.

In other words, The Law of Retail Gravitation is a concept in geography and economics that explains how people decide where to shop. It says that shoppers are attracted to bigger shopping centers that are closer to their homes.

This idea helps explain why some shopping centers become very popular and draw customers from a larger area, while smaller ones mainly serve the local neighborhood. It's like a "gravitational pull" that bigger shopping centers have on shoppers, bringing them in from farther distances.

For businesses and city planners, understanding the Law of Retail Gravitation can help them decide where to build new stores or malls. They can use this knowledge to create shopping areas that are convenient for people and attract a larger number of customers.

Credit to William J. Reilly and Historical Context: The Law of Retail Gravitation is credited to American economist William J. Reilly. Reilly developed this concept in the early 20th century as a response to the challenges faced by small retailers in competing with larger, more centralized retail centers. Reilly's work focused on understanding the forces that shape retail trade patterns and the factors influencing consumer choices in selecting retail locations.

Example of the Law of Retail Gravitation in Real Life: An example of the Law of Retail Gravitation can be observed in the choice of consumers to travel to larger shopping malls or retail centers located at greater distances. Suppose there are two retail centers, Center A and Center B, with Center A being closer to a residential area. However, Center B offers a wider range of stores, competitive prices, and additional amenities. Consumers may be *more* likely to travel a greater distance to Center B, as it provides a more attractive and comprehensive shopping experience. This example demonstrates how consumer choices are influenced by the size, attractiveness, and convenience of retail centers, even if they require traveling a longer distance.

Importance of the Law of Retail Gravitation: The Law of Retail Gravitation holds significant importance in the field of retail management, urban planning, and consumer behavior:

1. Retail Site Selection: Understanding the Law of Retail Gravitation helps retailers in determining optimal site locations for their businesses. By considering factors such as consumer preferences, market attractiveness, and competitive advantages, retailers can strategically position themselves to attract and retain customers, increasing their chances of success.

2. Urban Planning and Development: The law has implications for urban planning and development. City planners and policymakers can utilize this principle to assess the distribution and accessibility of retail centers within a city or region. It aids in determining the appropriate location and size of retail centers, promoting balanced retail development and ensuring equitable access to shopping opportunities.

3. Consumer Behavior Analysis: The law contributes to the understanding of consumer behavior and decision-making processes. By recognizing the factors that influence consumers' choices, such as distance, variety, price, and convenience, marketers and researchers can tailor their strategies to meet consumer needs and preferences more effectively.

4. Competitive Advantage: The law highlights the importance of competitiveness for retail centers. Retailers and shopping centers need to create an attractive shopping environment, offer a wide range of products and services, provide

competitive pricing, and ensure convenience to retain and attract customers. By understanding the factors that drive consumer choices, retailers can develop strategies to gain a competitive advantage in the market.

5. Retail Planning and Revitalization: The law aids in retail planning and the revitalization of retail areas. By understanding the forces of retail gravitation, planners can identify areas that require support and intervention, such as providing incentives for small retailers or implementing improvements to enhance the attractiveness of existing retail centers.

Conclusion: The Law of Retail Gravitation provides insights into consumer behavior and the dynamics of retail trade. This principle underscores the significance of distance and attractiveness in shaping consumer choices and preferences. Understanding the Law of Retail Gravitation allows retailers, urban planners, and policymakers to make informed decisions regarding retail site selection, urban development, and consumer-oriented strategies. By considering factors such as convenience, variety, and price, retailers can enhance their competitive position and provide an optimal shopping experience for consumers. The law serves as a valuable tool in understanding retail dynamics, guiding retail planning, and fostering sustainable retail development within communities.

6

PRINCIPLE OF LEAST EFFORT

The Principle of Least Effort, also known as the Law of Least Effort or the Path of Least Resistance, is a concept that explains human behavior and decision-making based on the tendency to minimize effort and maximize efficiency. This principle suggests that individuals are naturally inclined to choose the path or action that requires the least amount of energy, time, or cognitive effort.

Definition of the Principle of Least Effort: The Principle of Least Effort suggests that individuals have a natural inclination to choose the option that requires the least amount of effort to achieve a desired outcome. This principle assumes that humans have limited cognitive resources and seek to conserve them by opting for the most efficient or effortless course of action. It applies to various aspects of human behavior, including cognitive processes, problem-solving, and decision-making.

Imagine you have two ways to get to school: one is a long, hilly path, and the other is a shorter, flat road. The Principle of Least Effort suggests that most people would choose the shorter and easier road to save time and energy.

This principle applies to many aspects of life, not just physical tasks. For example, when solving problems or making decisions, people often opt for the simplest and most direct solution instead of a more complicated one.

The Principle of Least Effort helps us understand why certain patterns and behaviors emerge in society. It shows that people naturally seek efficiency and convenience in their actions, which can influence how cities are designed, how products are developed, and how we organize our daily lives.

Credit to George Kingsley Zipf and Historical Context: The Principle of Least Effort is attributed to linguist George Kingsley Zipf, who first proposed this concept in the field of linguistics during the mid-20th century. Zipf observed that individuals tend to use and choose words in a manner that minimizes effort. His work on language and communication patterns inspired the application of this principle to other areas of human behavior.

Example of the Principle of Least Effort in Real Life: An example of the Principle of Least Effort can be observed in everyday decision-making. Suppose you have multiple tasks to complete within a limited timeframe. In order to conserve energy and time, you may prioritize and choose the tasks that require the least amount of effort or offer the greatest efficiency. This could involve completing simpler tasks first, automating repetitive processes, or finding shortcuts to accomplish goals. By employing the Principle of Least Effort, individuals aim to achieve desired outcomes while minimizing unnecessary exertion.

Importance of the Principle of Least Effort: The Principle of Least Effort holds significant importance in understanding human behavior and decision-making:

1. Cognitive Efficiency: This principle acknowledges that human cognitive resources are limited. By recognizing the inclination to conserve mental energy, individuals can make more efficient use of their cognitive abilities, enhancing productivity and reducing decision fatigue.

2. Behavior Prediction: The Principle of Least Effort provides insights into predicting human behavior. By understanding individuals' preference for efficiency and the path of least resistance, researchers, marketers, and policymakers can anticipate and influence decision-making processes, facilitating desired outcomes.

3. Design and User Experience: This principle informs the design of products, services, and user interfaces. Designers can optimize user experiences by minimizing complexity, reducing cognitive load, and providing intuitive and effortless interactions. Applying the Principle of Least Effort enhances usability, satisfaction, and adoption rates.

4. Problem-Solving and Creativity: The Principle of Least Effort encourages individuals to seek efficient solutions to problems. By examining multiple options and considering the effort required, individuals can identify innovative approaches and streamline problem-solving processes.

5. Decision-Making and Behavioral Economics: The principle aligns with behavioral economics, which recognizes that

individuals often make choices based on factors such as effort, convenience, and the perceived cost-benefit ratio. Incorporating the Principle of Least Effort in decision-making models helps explain and predict human choices in various contexts.

Conclusion: The Principle of Least Effort offers insights into human behavior and decision-making processes. This principle highlights the natural inclination of individuals to choose the option that requires the least amount of effort to achieve desired outcomes. Understanding the Principle of Least Effort provides valuable perspectives on cognitive efficiency, behavior prediction, design, problem-solving, and decision-making. By considering this principle, individuals can optimize their cognitive resources, designers can enhance user experiences, and decision-makers can create environments that encourage efficient and effective choices. Recognizing the role of effort and efficiency in human behavior allows us to better understand ourselves and the choices we make in a complex world where resources and attention are limited.

7

THEORY OF HUMAN ECOLOGY

The Theory of Human Ecology is a multidisciplinary approach that examines the relationship between humans and their environment. This theory recognizes the reciprocal influence between human societies and their ecological surroundings, emphasizing the interconnectedness of social, cultural, and environmental factors.

Definition of the Theory of Human Ecology: The Theory of Human Ecology views humans as integral components of ecosystems, where their actions and behaviors shape and are influenced by their surrounding environment. It recognizes that human societies are influenced by social, cultural, economic, and political factors, as well as natural resources, climate, and other ecological elements. This theory emphasizes the interconnectedness and interdependence between humans and their environment, highlighting the reciprocal nature of the relationship.

Imagine you are studying a community and its surroundings like a scientist. Human ecology helps you see how people and their activities are shaped by the natural environment around them. For example, the type of jobs people have, the houses they build, and the food they eat can all be influenced by factors like climate, resources, and geography.

This theory also explores how humans, as a society, impact the environment in return. For instance, how we use land for agriculture or build cities can affect nature and other living things.

By understanding human ecology, we can make better decisions about how to live sustainably, taking care of both ourselves and the planet. It reminds us that we are connected to the natural world and that our actions can have a big impact on the environment and our well-being.

Credit to Amos Hawley and Historical Context: The Theory of Human Ecology is credited to sociologist Amos Hawley, who developed one definition of this concept in the mid-20th century. Hawley's work focused on the interaction between population dynamics and the environment, emphasizing the role of social structures, adaptation, and resource utilization. He integrated principles from sociology, biology, and geography to form a holistic framework for understanding human-environment relationships.

Example of the Theory of Human Ecology in Real Life: An example of the Theory of Human Ecology can be observed in the study of urbanization and its impacts on the environment. As populations grow and urban areas expand, human activities influence and are influenced by the surrounding ecological systems. This theory recognizes that urbanization affects not only social and economic aspects but also natural resources, biodiversity, air quality, and energy consumption. By applying the Theory of Human Ecology, researchers and policymakers can assess the environmental consequences of urbanization, develop strategies for sustainable urban planning, and promote the well-being of both humans and the environment.

Importance of the Theory of Human Ecology: The Theory of Human Ecology holds significant importance in understanding and addressing complex environmental and social challenges:

1. Sustainability and Resilience: This theory provides a framework for analyzing the interactions between human societies and the environment, fostering a more comprehensive understanding of sustainability and resilience. By recognizing the reciprocal relationship, it encourages the development of strategies that balance human needs with ecological integrity, ensuring the long-term well-being of both.

2. Environmental Planning and Management: The Theory of Human Ecology informs environmental planning and management practices. By considering the social, economic, and cultural dimensions alongside ecological factors, planners can design and implement policies that promote sustainable resource management, land use planning, and conservation efforts.

3. Human Well-being: The theory highlights the importance of considering human well-being in environmental decision-making. It recognizes that the quality of the environment significantly impacts human health, social cohesion, and overall quality of life. Understanding the Theory of Human Ecology guides efforts to create environments that support human well-being while protecting and enhancing ecological systems.

4. Social Justice and Equity: The theory addresses social justice and equity concerns in relation to the environment. It

recognizes that environmental degradation and resource exploitation disproportionately affect marginalized communities. Applying the Theory of Human Ecology enables policymakers to consider the distributional impacts of environmental decisions and work towards more equitable and inclusive outcomes.

5. Interdisciplinary Collaboration: The Theory of Human Ecology promotes interdisciplinary collaboration by integrating insights from various fields, such as sociology, geography, anthropology, and ecology. This multidisciplinary approach encourages researchers, practitioners, and policymakers to collaborate and address complex environmental challenges that require holistic and integrated solutions.

Conclusion: The Theory of Human Ecology as developed by Amos Hawley, provides a valuable framework for understanding the interaction between humans and their environment. This theory recognizes the reciprocal relationship and interdependence between human societies and ecological systems. Understanding the Theory of Human Ecology is essential for promoting sustainable development, addressing environmental challenges, and ensuring the well-being of both humans and the environment. By integrating social, cultural, economic, and ecological perspectives, this theory facilitates comprehensive and integrated approaches to environmental planning, resource management, and policy formulation. Embracing the principles of the Theory of Human Ecology allows us to navigate the complexities of human-environment interactions and work towards a more sustainable and resilient future.

8

THEORY OF SPATIAL DIFFUSION

The Theory of Spatial Diffusion is a concept that explains the spread and adoption of innovations, ideas, and cultural traits across geographic space. It examines how these elements diffuse from their origin to other locations, influencing social, economic, and cultural landscapes.

Definition of the Theory of Spatial Diffusion: The Theory of Spatial Diffusion examines the process by which innovations, ideas, or cultural traits spread across space, affecting different geographic locations and populations. It explores the mechanisms, patterns, and factors that influence the diffusion process. This theory recognizes that the diffusion of innovations can occur through various channels, such as migration, communication networks, trade routes, and cultural interactions.

Imagine you have a new and exciting dance move that you show to your friends at school. They think it's cool and start doing it too. Soon, more and more people in your school start doing the dance move, and eventually, it becomes popular across the entire town. This process of the dance move spreading from one person to another and from one place to another is an example of spatial diffusion.

The theory helps us understand how things like languages, technologies, fashion trends, or even diseases can move from one community to another, sometimes across countries or continents. There are different ways this can happen, like through trade, migration, communication, or simply people observing and copying each other.

By studying spatial diffusion, we can see how cultures interact and influence each other, how innovations and knowledge spread, and how our world becomes more connected through the sharing of ideas and practices.

Credit to Hagerstrand, and Historical Context: The Theory of Spatial Diffusion has been developed and expanded upon by various scholars over time. Torsten Hagerstrand, a Swedish geographer, made significant contributions to understanding the mechanisms and patterns of spatial diffusion. Additionally, Everett Rogers' work on the Diffusion of Innovations, particularly the concept of the diffusion curve and the role of opinion leaders, has influenced the theory's development.

Example of the Theory of Spatial Diffusion in Real Life: An example of the Theory of Spatial Diffusion can be observed in the spread of mobile phone technology. Believe it or not, the initial adoption of mobile phones was limited to a few geographical regions or countries. However, as the technology improved, costs decreased, and the benefits became evident, the diffusion process accelerated. Mobile phones gradually spread from the early adopters to the early majority and, eventually, to the late majority and laggards. This diffusion was facilitated by communication networks, trade routes, and cultural interactions. Today, mobile phones have become a ubiquitous technology worldwide. This example illustrates how innovations can diffuse across space, transforming societies and influencing social dynamics.

Importance of the Theory of Spatial Diffusion: The Theory of Spatial Diffusion holds significant importance in understanding societal and cultural change, as well as innovation adoption:

1. Social and Cultural Change: The theory provides insights into the spread of cultural practices, beliefs, and innovations, influencing social dynamics and cultural landscapes. It helps researchers understand the mechanisms and patterns through which changes occur within societies, shaping norms, values, and behaviors.

2. Innovation Diffusion: The theory is crucial in studying the adoption and spread of innovations. By examining the factors that affect the diffusion process, such as communication channels, social networks, and cultural contexts, researchers can identify strategies to promote innovation adoption and diffusion, ultimately driving social and economic progress.

3. Globalization and Cultural Exchange: The theory contributes to the understanding of globalization and cultural exchange processes. It reveals how ideas, practices, and technologies traverse borders and influence diverse societies. Understanding spatial diffusion aids in the analysis of cultural interactions, hybridization, and the formation of global networks.

4. Policy and Planning: The theory informs policy and planning efforts by considering the spatial dimension of innovation diffusion. Policymakers can utilize this understanding to identify areas with limited access to innovations and develop strategies to promote equitable diffusion. Spatial

diffusion analysis assists in identifying potential barriers and devising targeted interventions to bridge gaps and reduce disparities.

5. Marketing and Business Strategies: The theory has implications for marketing and business strategies. It helps businesses identify target markets, understand consumer behavior, and develop effective marketing campaigns. By understanding the diffusion process, businesses can design strategies to reach early adopters and influence the broader market.

Conclusion: The Theory of Spatial Diffusion provides valuable insights into the spread and adoption of innovations, ideas, and cultural traits across geographic space. It recognizes the importance of spatial processes and social dynamics in shaping diffusion patterns. Understanding the Theory of Spatial Diffusion aids in understanding societal and cultural change, innovation adoption, globalization, and policy planning. By comprehending the mechanisms and factors that influence spatial diffusion, researchers, policymakers, and businesses can better navigate the complexities of diffusion processes and leverage them to drive positive social, economic, and cultural transformations.

PART 2

PHYSICS

Physics

Physics is a branch of science that explores the fundamental principles governing our universe. It seeks to understand and explain the behavior of matter and energy across all scales, from the smallest subatomic particles to vast cosmological phenomena. It is a field characterized by quantitative reasoning, rigorous experimentation, and conceptual understanding. Physics addresses questions such as: What are the basic constituents of the universe? How do they interact with each other? And what are the laws that govern these interactions?

The study of physics is profoundly important for many reasons. First, it forms the foundation for other scientific disciplines, including chemistry, biology, and earth science, which in turn shape our understanding of the natural world. Second, the laws and principles uncovered through physics have been instrumental in developing countless technologies that underpin our modern life, from electricity and semiconductors to lasers and magnetic resonance imaging. Third, physics fosters a problem-solving mindset, enabling the development of analytical and critical thinking skills, which are highly valuable in a range of professions. Finally, at a deeper level, studying physics helps us grasp our place in the universe, enriching our worldview and inspiring us to seek answers to the most profound questions of existence.

9

Law of Universal Gravitation

The law of universal gravitation is a fundamental principle that governs the behavior of celestial bodies, you know, stars, planets, things like that and shapes the dynamics of our universe. Proposed by Sir Isaac Newton, this law has revolutionized our understanding of gravity and played a pivotal role in advancing various fields of science.

Definition of the law of universal gravitation: The law of universal gravitation states that every particle of matter in the universe attracts every other particle with a force that is directly proportional to the product of their masses and inversely proportional to the square of the distance between them. In simpler terms, any two objects with mass will experience an attractive force pulling them toward each other. This force of gravity depends on the masses of the objects and the distance between them.

Imagine you are holding a ball in your hand. The Earth's gravity pulls the ball towards the ground, and that's why you feel its weight. But did you know that the ball also pulls the Earth? According to the Law of Universal Gravitation, every object with mass in the universe attracts every other object with mass. The strength of this attraction depends on how massive the objects are and how far apart they are from each other.

This simple yet fundamental law helps scientists understand the motion of celestial bodies, the orbits of planets, and even things like the tides in our oceans. It is one of the cornerstones of modern physics and has greatly contributed to our understanding of the universe and how it works.

Credit to Sir Isaac Newton: The credit for formulating the law of universal gravitation goes to the English physicist and mathematician, Sir Isaac Newton. In his groundbreaking work "Philosophiæ Naturalis Principia Mathematica" (Mathematical Principles of Natural Philosophy), published in 1687, Newton presented his laws of motion and the law of universal gravitation. Newton's profound insights and mathematical calculations not only explained the motion of the planets but also offered a cohesive framework to understand the fundamental force that governs the universe.

Example of the Law in Real Life: An illustrative example of the law of universal gravitation in real life is the motion of Earth's moon around our planet. The moon, despite being approximately 384,400 kilometers away from Earth, remains in a stable orbit due to the force of gravity. According to the law, Earth's mass and the moon's mass create a gravitational attraction between them. The moon continuously falls toward Earth due to this gravitational pull, but its tangential velocity keeps it in a perpetual orbit, resulting in a delicate equilibrium between the gravitational force and the moon's centripetal force.

Importance of the Law of Universal Gravitation: The law of universal gravitation is of paramount importance for various reasons:

1. Understanding Celestial Mechanics: This law provides a foundation for understanding the behavior of planets, stars, galaxies, and other celestial objects. By applying the law of universal gravitation, astronomers can predict the motion of planets, the formation of solar systems, and the dynamics of astronomical phenomena.

2. Satellite and Spacecraft Trajectories: The law of universal gravitation enables scientists to calculate and design the trajectories of satellites and spacecraft, ensuring precise navigation and orbital stability. This knowledge is crucial for various applications, such as telecommunications, weather forecasting, and global positioning systems (GPS).

3. Planetary Science and Exploration: The law of universal gravitation aids in the study and exploration of other planets and celestial bodies. By understanding the gravitational interactions between these objects, scientists can analyze the composition, atmosphere, and geological features of planets, moons, asteroids, and comets.

4. Fundamental Scientific Understanding: Newton's law of universal gravitation laid the groundwork for the development of classical physics and marked a significant step forward in our comprehension of the fundamental forces of nature. It set the stage for subsequent scientific advancements, including Albert Einstein's theory of general relativity.

Conclusion: The law of universal gravitation, continues to shape our understanding of the universe. Its ability to explain the motions of celestial bodies, predict orbits, and guide space exploration highlights its immense importance. This law has transformed our perception of gravity and remains a cornerstone of modern physics, leaving an indelible mark on the scientific community and our quest for knowledge about the cosmos.

10

THEORY OF RELATIVITY

The theory of relativity, one of the most profound scientific theories of the 20th century, transformed our understanding of space, time, and gravity. This theory revolutionized the field of physics and challenged long-held Newtonian notions. It consists of two parts: the Special Theory of Relativity and the General Theory of Relativity.

The Special Theory of Relativity deals with how things move when they are going very fast. Think of the speed of light. It tells us that time can slow down, and objects can become shorter as they move faster. It also shows that the laws of physics are the same for everyone, no matter how fast they are moving. This theory helped us understand how the universe works on a very small scale.

The General Theory of Relativity, on the other hand, is about gravity and how it affects the shape of space and the flow of time. It explains that massive objects, like stars and planets, create a "dent" in space-time, which causes other objects nearby to be pulled towards them. This theory revolutionized our understanding of gravity and how it works on a large scale, like in our solar system or galaxies.

Definition of the Theory of Relativity: The theory of relativity encompasses two major pillars, as mentioned previously: the special theory of relativity and the general theory of relativity. The special theory, proposed in 1905, established that the laws of physics are the same for all observers, regardless of their relative motion. It introduced the concept of the speed of light as an absolute cosmic speed limit and led to the famous equation $E=mc^2$, which relates energy (E) to mass (m) and the speed of light (c). The general theory of relativity, came later, in 1915, and expanded on these ideas and provided a new understanding of gravity by considering the curvature of spacetime.

Credit to Albert Einstein: Albert Einstein, the physicist and mathematician, is credited with the creation of the theory of relativity. In 1905, at the age of 26, Einstein published his groundbreaking paper on the special theory of relativity, challenging the classical Newtonian notions of absolute space and time. Over the next decade, he continued his work and developed the general theory of relativity, introducing the revolutionary concept of gravity as the curvature of spacetime caused by the presence of mass and energy.

Example of the Theory of Relativity in Real Life: An example of the theory of relativity in real life is the phenomenon known as time dilation. According to the special theory of relativity, time is not absolute but depends on the relative motion between observers. When objects or individuals move at speeds close to the speed of light, time slows down for them relative to a stationary observer. This effect has been experimentally confirmed, notably in the context of particle accelerators and satellite navigation systems. GPS satellites, for instance, rely on precise timekeeping, and their accurate functioning requires compensating for the time dilation effects due to their high orbital speeds.

Importance of the Theory of Relativity: The theory of relativity holds immense importance for several reasons:

1. Redefining Space, Time, and Gravity: The theory of relativity revolutionized our understanding of space, time, and gravity. By challenging the traditional Newtonian framework, it provided a more comprehensive and accurate description of the universe, paving the way for further scientific advancements.

2. GPS and Satellite Navigation: The theory of relativity plays a crucial role in the functioning of GPS and satellite navigation systems. The precise calculations and adjustments made based on the theory ensure accurate positioning and timing, allowing for reliable global navigation.

3. Cosmology and the Big Bang Theory: The theory of relativity provides the theoretical foundation for modern cosmology. It helps us understand the evolution of the universe, including the Big Bang theory, the expansion of space, and the formation of galaxies, stars, and planets.

4. Fundamental Physics and Particle Accelerators: The theory of relativity is a fundamental component of modern physics. It underpins our understanding of particle physics, quantum field theory, and high-energy phenomena. Particle accelerators, such as the Large Hadron Collider (LHC), rely on the principles of relativity to investigate the fundamental building blocks of matter and unravel the mysteries of the universe.

Conclusion: The theory of relativity, stands as one of the most remarkable scientific achievements in history. Its transformative ideas have redefined our concepts of space, time, and gravity, revolutionizing physics and our understanding of the universe. From everyday technologies like GPS to the exploration of the cosmos and the pursuit of fundamental physics, the theory of relativity continues to shape scientific progress, inspiring new discoveries and challenging us to delve deeper into the mysteries of the cosmos.

11

QUANTUM MECHANICS

Quantum mechanics, a cornerstone of modern physics, unveils the fascinating and often perplexing behavior of particles at the subatomic level. This revolutionary theory has transformed our understanding of the microscopic world and shattered classical notions of determinism.

Definition of Quantum Mechanics: Quantum mechanics is a branch of physics that describes the behavior of matter and energy at the smallest scales. It provides a mathematical framework to understand the wave-particle duality of matter and the probabilistic nature of physical phenomena. Unlike classical mechanics, which follows deterministic principles, quantum mechanics introduces inherent uncertainty and incorporates wave-like properties, such as superposition and entanglement, to explain the behavior of particles.

In more simple terms, it helps us understand the behavior of very tiny particles, like atoms and subatomic particles.

Imagine you have a tiny toy car, and you want to know exactly where it is and how fast it's moving. Classical physics would say you can measure its position and speed precisely. But with quantum mechanics, things are a bit different. It tells us that at the tiny scale of atoms and particles, things can be a bit uncertain and unpredictable.

In the quantum world, particles can behave like both particles and waves, and we can't precisely measure both their position and speed at the same time. Instead, we use probabilities to describe where they are likely to be and how fast they might be moving. This is a bit like predicting the chances of flipping a coin and getting heads or tails.

Quantum mechanics has some odd and counterintuitive aspects, like particles being in multiple places at once or communicating instantaneously over large distances. However, it's a remarkably accurate theory that has been tested and confirmed through countless experiments.

Quantum mechanics has led to many technological advancements, including computers, lasers, and even the internet. It has also opened up new frontiers in science, revealing the mysterious and wondrous nature of the tiniest building blocks of our universe.

Credit to Founders of Quantum Mechanics: Several pioneering scientists contributed to the development of quantum mechanics. In the early 20th century, Max Planck introduced the concept of quantized energy, leading to the birth of quantum theory. Later, Albert Einstein, Niels Bohr, Erwin Schrödinger, Werner Heisenberg, and others made significant contributions to formulating the principles and mathematical foundations of quantum mechanics.

Example of Quantum Mechanics in Real Life: An example of quantum mechanics in real life is the phenomenon of quantum computing. Quantum computers harness the principles of quantum mechanics, such as superposition and entanglement, to perform computations that surpass the capabilities of classical computers. By utilizing quantum bits, or qubits, which can exist in multiple states simultaneously, quantum computers have the potential to solve complex problems more efficiently, including cryptography, optimization, and simulations of quantum systems.

Importance of Quantum Mechanics: The idea of quantum mechanics carries immense importance for several reasons:

1. Understanding Subatomic Physics: Quantum mechanics provides a framework to comprehend the behavior of particles at the quantum level. It explains phenomena such as particle-wave duality, the quantization of energy, and the probabilistic nature of measurements. This knowledge is vital for studying atomic and subatomic systems and advancing fields like quantum chemistry and solid-state physics.

2. Technological Advances: Quantum mechanics has enabled remarkable technological advancements. Applications such as transistors, lasers, magnetic resonance imaging (MRI), and semiconductor devices rely on quantum principles. Furthermore, ongoing research in quantum technologies, including quantum computing, quantum cryptography, and quantum sensing, holds the promise of transforming various industries and enhancing computational power, security, and sensing capabilities.

3. Fundamental Physics: Quantum mechanics plays a crucial role in our understanding of fundamental physics. It forms the foundation of quantum field theory, which unifies quantum mechanics with special relativity and describes the fundamental forces and particles of nature. Moreover,

quantum mechanics provides insights into the early universe, the behavior of black holes, and the mysterious phenomenon of quantum entanglement.

4. Philosophical Implications: Quantum mechanics challenges our traditional notions of reality and determinism. The theory's probabilistic nature and the observer's role in measurement have sparked philosophical debates about the nature of reality, consciousness, and the limits of human knowledge. These discussions continue to shape the intersection of physics, philosophy, and metaphysics.

Conclusion: Quantum mechanics has transformed our understanding of the microscopic world. By embracing the probabilistic nature of matter and energy, quantum mechanics has paved the way for technological advancements and deepened our knowledge of fundamental physics. From quantum computing to advanced materials and philosophical inquiries, the impact of quantum mechanics extends across disciplines, driving innovation and unraveling the mysteries of the subatomic realm.

12

LAW OF LARGE NUMBERS

The law of large numbers is a fundamental principle in statistics that provides insights into the behavior of random variables and the stability of their averages. This powerful law enables us to make reliable predictions and draw meaningful conclusions from large sets of data.

Definition of the Law of Large Numbers: The law of large numbers states that as the sample size increases, the average of a set of independent and identically distributed random variables converges to the expected value of the underlying distribution. In simpler terms, it suggests that as we collect more data, the observed outcomes tend to align with the expected or theoretical probabilities.

Imagine you have a fair six-sided die, and you roll it many times, recording the results. As you roll the die more and more times, you'll notice that the average number you get from all the rolls gets closer and closer to the expected average of 3.5 (the sum of all the numbers on the die divided by the total number of sides).

The Law of Large Numbers tells us that as we increase the number of trials or events, the average outcome approaches the true or expected probability. This means that if we were to roll the die an infinite number of times, the average would be exactly 3.5.

This principle is widely used in statistics and probability, and it helps us make more accurate predictions and estimates when dealing with random events. It assures us that, in the long run, the observed outcomes will converge to the theoretical probabilities, providing a foundation for reliable statistical analysis.

Credit to Jacob Bernoulli: The law of large numbers is credited to Swiss mathematician Jacob Bernoulli, who first formulated the concept in the 17th century. In his work "Ars Conjectandi" (The Art of Conjecturing), published posthumously in 1713, Bernoulli presented his investigations into probability theory. He described the law of large numbers as a fundamental principle for understanding the stability and predictability of random events.

Example of the Law of Large Numbers in Real Life: An example of the law of large numbers is the application of polling in elections. Polling organizations sample a relatively small subset of the population to gauge public opinion. By applying the law of large numbers, they assume that as the sample size increases, the poll results will converge to the true population opinion. This statistical principle allows pollsters to make reliable predictions about election outcomes based on representative samples.

Importance of the Law of Large Numbers: The law of large numbers holds significant importance for several reasons:

1. Statistical Stability: The law of large numbers provides a foundation for statistical inference by demonstrating that as the sample size increases, random fluctuations tend to

average out, resulting in more stable and reliable estimates. This stability allows researchers to draw valid conclusions from data and make informed decisions based on statistical analysis.

2. Risk Assessment and Insurance: The law of large numbers is crucial in the insurance industry. Insurers rely on large pools of policyholders to spread the risk of potential claims. By pooling a large number of individuals with similar risks, insurance companies can accurately predict the frequency and severity of claims, ensuring the financial stability of their operations.

3. Financial Markets: The law of large numbers is applicable to financial markets as well. Investors often rely on the principle that as the number of trades or investments increases, the average returns will converge to the expected returns. This principle helps guide investment strategies and risk management decisions in various financial instruments.

4. Predictive Analytics: The law of large numbers underpins the field of predictive analytics. By analyzing vast amounts of historical data, organizations can identify patterns, trends, and relationships that can be used to predict future outcomes. This knowledge has applications in areas such as marketing, customer behavior analysis, and demand forecasting.

Conclusion: The law of large numbers serves as a cornerstone of statistical theory and practice. Its ability to provide stability and predictability to random events has wide-ranging applications in fields such as polling, insurance, finance, and predictive analytics. By understanding the behavior of averages in large data sets, researchers and decision-makers can make informed choices, assess risks, and gain valuable insights into the world of uncertainty. The law of large numbers continues to shape the way we analyze data and make sense of the complexities of chance and probability.

13

THE CENTRAL LIMIT THEOREM

The Central Limit Theorem is a fundamental concept in statistics that explains the behavior of the sum or average of a large number of independent and identically distributed random variables. It provides valuable insights into the distribution of sample means and enables statisticians to make inferences about populations.

Definition of the Central Limit Theorem: The Central Limit Theorem states that when independent random variables are added or averaged, regardless of the underlying distribution, the resulting sum or average tends to follow a normal distribution* as the sample size increases. This remarkable theorem establishes the normal distribution as a robust and ubiquitous phenomenon, allowing statisticians to make probabilistic statements about sample means.

*A quick note about normal distribution: A normal distribution, also known as a bell curve, is a common and symmetrical pattern in statistics that describes how data tends to cluster around an average value, with fewer occurrences of extreme values on either side. It forms a smooth, bell-shaped curve when graphed.

In simple terms, imagine you are measuring the heights of a large group of people. If you plot these heights on a graph, you'll notice that most people will be around the average height, and as you move away from the average, there will be fewer and fewer individuals. The graph of this data will resemble a bell-shaped curve, and this pattern is what we call a normal distribution.

The Central Limit Theorem is a powerful concept in statistics that tells us how the averages of many random samples tend to follow a particular pattern, regardless of the original data's distribution.

Imagine you have a big jar of colorful candies, and you want to know the average number of candies per person in a group. To find out, you randomly pick a few people, count the candies they have, and calculate the average. Now, you repeat this process with different random groups of people and keep calculating their averages.

The Central Limit Theorem says that as you take more and more random samples and calculate their averages, the distribution of these averages will start to look like a smooth and bell-shaped curve, called a "normal distribution." This curve is also known as a "bell curve" because it looks like a symmetrical bell.

What's remarkable is that even if the original number of candies in the jar didn't follow a bell-shaped curve, the averages of the random samples still tend to form this pattern. This property makes the Central Limit Theorem very useful in statistics because it allows us to make predictions about the averages, even when we don't know the exact distribution of the original data.

The Central Limit Theorem is the reason why the bell curve is so common in statistics and why it appears in various fields, from social sciences to natural sciences and everything in between. It helps us make sense of data, understand patterns, and make more accurate predictions in a wide range of situations.

Credit to Abraham de Moivre: The Central Limit Theorem is commonly attributed to Abraham de Moivre, an 18th-century French mathematician. In 1733, de Moivre published "The Doctrine of Chances," where he presented the theorem in the context of binomial distributions. Although he did not provide a formal proof, de Moivre laid the groundwork for subsequent mathematicians to establish the theorem's validity.

Example of the Central Limit Theorem in Real Life: An example of the Central Limit Theorem in real life is seen in opinion polls. Pollsters often collect data from a sample of individuals to estimate the proportion of a population that holds a particular opinion. By applying the Central Limit Theorem, they can assume that as the sample size increases, the distribution of sample proportions will become approximately normal. This allows them to calculate confidence intervals and make statements about the likely range of population proportions.

Importance of the Central Limit Theorem: The Central Limit Theorem holds significant importance for several reasons:

1. Statistical Inference: The Central Limit Theorem is the cornerstone of statistical inference. It allows statisticians to make robust inferences about population parameters based on sample statistics. By assuming a normal distribution for sample means, they can construct confidence intervals, perform hypothesis tests, and draw conclusions about the pop-

ulation.

2. Reliability of Sample Means: The Central Limit Theorem demonstrates that sample means tend to be more stable and less affected by random fluctuations compared to individual observations. This property makes sample means valuable and reliable estimators of population means. It allows researchers to obtain more precise estimates by collecting larger samples.

3. Hypothesis Testing: The Central Limit Theorem plays a crucial role in hypothesis testing. It provides the foundation for constructing test statistics, such as Z-scores and T-scores, which are used to compare sample means with hypothesized population means. This enables researchers to evaluate the significance of observed differences and make informed decisions about the validity of hypotheses.

4. Quality Control and Process Improvement: The Central Limit Theorem is instrumental in quality control and process improvement methodologies. By collecting and analyzing samples, practitioners can monitor and control various process parameters, such as product dimensions or service response times. The theorem's application allows them to determine acceptable ranges and identify process deviations.

Conclusion: The Central Limit Theorem is a cornerstone of statistical theory and practice. Its ability to establish the normal distribution as a universal phenomenon for sample means has far-reaching implications in statistical inference, hypothesis testing, and quality control.

By understanding the behavior of averages and the convergence to normality, researchers and practitioners can make reliable estimates, perform rigorous statistical analyses, and make informed decisions based on data. The Central Limit Theorem continues to shape the field of statistics and plays a vital role in numerous applications across various domains.

14

THE PRINCIPLE OF LEAST ACTION

The Principle of Least Action is a fundamental concept in physics that provides a powerful framework for understanding the behavior of dynamic systems. It describes how objects in the universe move and behave and suggests that the way things move or change over time is determined by trying to take the "easiest" or "quickest" path possible.

Definition of the Principle of Least Action: The Principle of Least Action asserts that the path taken by a physical system between two points in time is the one that minimizes the action. Action is defined as the integral of the difference between a system's kinetic and potential energy over a given time interval. This principle suggests that natural processes follow a path of least resistance or effort, optimizing the energy balance of the system.

Imagine you are playing a game where you have to get from one point to another on a game board. The Principle of Least Action tells you to think carefully about how you move your game piece so that you spend the least effort or time to reach your destination. You'll choose the path that gets you there with the least amount of steps or movements.

In the real world, objects, like planets, particles, or light, also follow the Principle of Least Action. They move in a way that minimizes the "action," which is a specific mathematical quantity representing the effort needed to make a change in their motion.

This principle is essential because it helps us understand the motion of objects in the universe, from tiny particles to massive planets. It is used in various areas of physics, such as classical mechanics and quantum mechanics, to describe and predict how things move and interact with each other in the most efficient and natural way possible.

Credit to Pierre Louis Maupertuis: The Principle of Least Action is commonly attributed to Pierre Louis Maupertuis, a French mathematician and astronomer of the 18th century. In 1744, Maupertuis published the principle in his work "Essai de cosmologie" (Essay on Cosmology). Although Maupertuis faced some controversies and debates regarding his formulation, his contribution to the development of this principle is widely recognized.

Example of the Principle of Least Action in Real Life: An example of the Principle of Least Action in real life is the path taken by light when traveling between two points. According to the principle, light follows the path that minimizes the action, known as the path of least time. When light travels through different media with varying refractive indices, such as air and water, it adjusts its path to minimize the time taken to travel between two points. This behavior can be observed when light appears to bend when passing through a prism or when witnessing the formation of a rainbow.

Importance of the Principle of Least Action: The Principle of Least Action holds significant importance for several reasons:

1. Fundamental Principles of Physics: The Principle of Least Action is deeply connected to fundamental principles of physics, such as Newton's laws of motion and Hamilton's principle. It provides a powerful mathematical framework for describing and predicting the behavior of a wide range of physical systems, from classical mechanics to quantum mechanics.

2. Variational Principles: The principle is part of a broader class of variational principles, which encompass other fundamental principles, including Fermat's principle of least time for light and Hamilton's principle of stationary action in classical mechanics. These principles provide concise and elegant formulations of physical laws, enabling scientists to derive equations of motion and uncover fundamental relationships.

3. Optimization and Efficiency: The Principle of Least Action embodies the concept of optimization and efficiency in physical systems. By minimizing the action, natural processes seek paths that conserve energy and effort, resulting in more efficient and stable configurations. This principle allows us to understand the dynamics of various phenomena, ranging from celestial motion to particle interactions.

4. Quantum Field Theory and Fundamental Interactions: The Principle of Least Action plays a central role in quantum field theory, the framework that describes the behavior of elementary particles and their interactions. By applying the principle to quantum fields, physicists can derive the equations of motion for particles, predict their behavior, and

study the fundamental forces of nature.

Conclusion: The Principle of Least Action is a fundamental principle in physics that provides a deep understanding of the dynamics of physical systems. By minimizing the action, natural processes seek paths of least resistance, optimizing energy balance and efficiency. From the path of light to the behavior of particles in quantum field theory, this principle illuminates the underlying principles of the universe and guides our understanding of the fundamental laws of nature. The Principle of Least Action continues to shape the field of physics and serves as a cornerstone for advancing our knowledge of the physical world.

15

THE BIG BANG THEORY

The Big Bang Theory is a widely accepted scientific model that explains the origin and evolution of the universe. It postulates that the universe began as a singularity and has been expanding ever since.

Definition of the Big Bang Theory: The Big Bang Theory imagines that the universe originated from a singular point of infinite density and temperature, often referred to as a singularity. Approximately 13.8 billion years ago, this singularity underwent a rapid expansion known as the Big Bang. The theory suggests that all matter, energy, space, and time originated from this primordial event, leading to the formation of galaxies, stars, and all cosmic structures.

In other words, the Big Bang Theory is an explanation for how our universe began. It suggests that the universe started from an incredibly hot and dense state billions of years ago.

Imagine the universe as a balloon that is getting bigger and bigger. If you go backward in time, the balloon gets smaller and smaller until it reaches a point where it was very tiny, almost like a single point. This is what scientists call the "singularity," the starting point of the Big Bang.

At the moment of the Big Bang, all the matter and energy in the universe were concentrated in this extremely hot and dense singularity. Then, in an instant, it began expanding and cooling down, rapidly filling the space with particles, atoms, and eventually stars and galaxies.

The Big Bang Theory helps us understand the origin of the universe and how everything we see today, including stars, planets, and even ourselves, can be traced back to that initial explosive event. It is supported by various observations and evidence, such as the cosmic microwave background radiation and the expansion of the universe, making it the most widely accepted explanation for the beginning of our cosmos.

Credit to Georges Lemaître and Edwin Hubble: The Big Bang Theory has its roots in the works of several scientists, but two individuals stand out for their contributions. Belgian physicist and priest Georges Lemaître first proposed the theory, envisioning an expanding universe arising from an initial state of extreme density. In the 1920s, American astronomer Edwin Hubble provided observational evidence for the expansion of the universe, reinforcing Lemaître's theoretical framework.

Example of the Big Bang Theory in Real Life: An example of the Big Bang Theory in real life is the observed cosmic microwave background radiation (CMB). As the universe expanded and cooled after the initial Big Bang, the release of intense energy resulted in a primordial glow that permeates the entire cosmos. This relic radiation, known as CMB, was discovered in 1965 and provides compelling evidence for the Big Bang Theory. The CMB exhibits a nearly uniform distribution of microwave radiation across the sky, supporting the idea of a hot, dense early universe.

Importance of the Big Bang Theory: The Big Bang Theory holds significant importance for several reasons:

1. Explaining the Origin of the Universe: The Big Bang Theory provides a comprehensive explanation for the origin of the universe. By postulating an initial singularity and subsequent expansion, it offers insights into the early moments of cosmic evolution. It addresses questions about the birth of space, time, and matter and provides a coherent framework for understanding the vastness of the cosmos.

2. Expanding Our Knowledge of Cosmology: The Big Bang Theory has revolutionized our understanding of the universe's evolution. It has allowed scientists to probe the early stages of the universe, study the formation of galaxies and cosmic structures, and investigate the distribution of matter and energy throughout space. It continues to shape the field of cosmology and guides our exploration of the cosmos.

3. Supporting Astronomical Observations: The Big Bang Theory is consistent with numerous observational and experimental findings. It aligns with Hubble's discovery of the universe's expansion, the abundance of light elements in the cosmos, and the cosmic microwave background radiation. The theory's predictions and compatibility with observational data provide strong evidence for its validity.

4. Unifying Fundamental Physics: The Big Bang Theory connects the realms of cosmology and fundamental physics. It highlights the interplay between gravity, quantum physics, and the behavior of matter and energy in extreme conditions.

It has inspired further investigations into the fundamental forces and particles of the universe and serves as a bridge between the macroscopic and microscopic realms of physics.

Conclusion: The Big Bang Theory offers a comprehensive explanation for the origin and evolution of the universe. It unravels the birth of space, time, and matter and provides a framework for understanding the vastness of the cosmos. By aligning with observational data and unifying fundamental physics, the Big Bang Theory has reshaped our understanding of the universe and continues to inspire exploration, discovery, and the quest for knowledge about our cosmic origins.

16

Newton's Laws

Newton's three laws of motion are essential and interconnected principles that form the foundation of classical mechanics and our understanding of how objects move and interact in the physical world. They are discussed together in this chapter because collectively they provide a comprehensive framework for explaining a wide range of physical phenomena. Together, Newton's three laws provide a unified and comprehensive understanding of the fundamental principles governing motion and force in the universe. They serve as the building blocks for many other areas of physics and engineering, and their principles continue to be used in countless practical applications, from designing vehicles and structures to understanding celestial mechanics and space exploration. The discussion of these laws together allows us to see the interconnectedness of the physical world and how these laws work in harmony to describe the behavior of objects and the forces that act upon them.

Newton's First law, Law of Inertia

Newton's First Law, also known as the law of inertia, is a fundamental principle in physics that describes the behavior of objects at rest or in uniform motion. This law forms the basis for understanding the concept of inertia, which governs the tendency of objects to maintain their state of motion.

Definition of Newton's First Law: Newton's First Law states that an object at rest remains at rest, and an object in motion continues to move with a constant velocity in a straight line unless acted upon by an external force. In simpler terms, an object will naturally resist changes in its motion, either maintaining its current state or requiring an external force to alter its velocity or direction.

Imagine you have a toy car on a flat, smooth surface, and you give it a push. The car moves forward and then slows down and stops. The Law of Inertia tells us that if no other force, like friction or a push, acts on the car, it will keep doing what it's doing—either staying at rest or moving at a constant speed in a straight line.

In even more simpler terms, the Law of Inertia states that objects tend to "keep doing what they are doing." If something is at rest, it will stay at rest until a force makes it move. If something is already moving, it will keep moving in a straight line at the same speed unless another force makes it stop or change its direction.

Credit to Sir Isaac Newton: Sir Isaac Newton, the renowned English physicist and mathematician, is credited with formulating Newton's First Law. This law, along with his other laws of motion, was presented in his groundbreaking work "Philosophiæ Naturalis Principia Mathematica" (Mathematical Principles of Natural Philosophy) published in 1687. Newton's laws of motion revolutionized the understanding of motion, gravity, and the fundamental principles of physics.

Example of Newton's First Law in Real Life: An example of Newton's First Law in real life is observed when a passenger in a moving vehicle experiences a sudden deceleration. If a car suddenly stops, the passengers inside continue moving forward due to their inertia. This is why it is crucial to wear seat belts as they exert a force on the passengers, restraining them and preventing them from continuing to move forward at the vehicle's original speed.

Importance of Newton's First Law: Newton's First Law holds significant importance for several reasons:

1. Foundation of Inertia: Newton's First Law establishes the concept of inertia, which is the resistance of an object to changes in its state of motion. It explains why objects tend to maintain their velocity and direction unless acted upon

by external forces. Understanding inertia is fundamental to comprehending the behavior of objects in motion and plays a vital role in various fields of science and engineering.

2. Everyday Mechanics: Newton's First Law has practical applications in everyday life. It helps explain phenomena such as why objects on a table remain at rest until acted upon by external forces, why it is challenging to start or stop a heavy object, or why passengers are pushed backward in a moving vehicle when it accelerates suddenly. This law provides insights into the mechanics of our surroundings.

3. Foundation for Newton's Laws of Motion: Newton's First Law serves as the foundation for his other two laws of motion. It establishes the concept of inertia, which is further expanded upon in the second law ($F = ma$) and leads to the understanding of action and reaction forces in the third law. The three laws collectively form the basis for classical mechanics and have broad applications in understanding the behavior of objects and systems.

4. Scientific and Engineering Applications: Newton's First Law is essential for scientific and engineering applications. It helps engineers design structures and vehicles that can withstand forces and resist changes in motion. It is also crucial for the development of technologies such as airbags in vehicles, which employ Newton's First Law to protect passengers by reducing the impact forces during collisions.

Conclusion: Newton's First Law lays the foundation for understanding inertia and the behavior of objects at rest or in motion. It explains why objects tend to maintain their state of motion and resist changes unless acted upon by external forces. This law has broad applications in various fields and is fundamental to comprehending the mechanics of our everyday world. Newton's First Law, alongside his other laws of motion, revolutionized our understanding of motion and remains a cornerstone of classical mechanics.

Newton's Second Law, Law of Force and Acceleration

Newton's second law is called the "Law of Acceleration" or the "Law of Motion." It is often referred to as simply "Newton's Second Law" and, is a fundamental principle in physics that establishes a quantitative relationship between the force applied to an object, its mass, and the resulting acceleration. This law provides insights into how external forces act upon objects, causing changes in their motion

Definition of Newton's Second Law: Newton's Second Law states that the acceleration of an object is directly proportional to the force applied to it and inversely proportional to its mass. Mathematically, it is expressed as $F = ma$, where F represents the force applied to the object, m represents its mass, and a represents the resulting acceleration. This law describes how a force acting on an object causes it to accelerate, with the magnitude of acceleration depending on the applied force and the object's mass.

Imagining the same toy car from the example in Newton's First Law and you push it with different amounts of force. If you push it gently, it moves slowly, but if you push it harder, it moves faster. Newton's Second Law tells us that the acceleration of the car (how fast it speeds up or slows down) is directly proportional to the force you apply to it. The more force you apply, the more the car accelerates.

The law also tells us that the acceleration depends on the mass of the object. If you push the same toy car and then a bigger and heavier toy truck with the same force, the car will accelerate more because it has less mass than the truck. This means that lighter objects are easier to accelerate than heavier ones.

Example of Newton's Second Law in Real Life: An example of Newton's Second Law in real life can be observed in a person pushing a grocery cart. When a person exerts a force on a cart, the cart experiences acceleration. The magnitude of the cart's acceleration depends on the force applied and its mass. A smaller cart will accelerate more quickly than a larger cart when the same force is applied since the smaller cart has less mass, following the relationship described by Newton's Second Law.

Importance of Newton's Second Law: Newton's Second Law holds significant importance for several reasons:

1. Quantitative Relationship: Newton's Second Law establishes a quantitative relationship between force, mass, and acceleration. It provides a mathematical framework to predict and calculate the resulting acceleration of an object when a force is applied. This relationship allows scientists and engineers to analyze and design systems involving forces, such as structures, vehicles, and machinery.

2. Understanding Dynamics: Newton's Second Law is essential for understanding the dynamics of objects in motion. It reveals how forces influence an object's acceleration, enabling predictions about how the object will respond to various forces. This understanding is crucial for analyzing and optimizing the performance of systems ranging from rockets and vehicles to athletic movements.

3. Practical Applications: Newton's Second Law finds practical applications in various fields. It guides the design of vehicles, considering factors such as engine power, mass distribution, and aerodynamics. It helps engineers optimize machines and equipment by analyzing the forces involved. The law is also crucial for designing safety features such as seat belts, airbags, and impact-resistant materials to protect individuals from excessive forces during accidents.

4. Bridge to Other Laws: Newton's Second Law serves as a bridge to other fundamental principles of physics. It connects to Newton's First Law (the law of inertia) and Third Law (the law of action and reaction), forming a comprehensive framework for understanding the behavior of objects and systems. It lays the groundwork for further exploration

in mechanics, such as rotational motion and gravitation.

Conclusion: Newton's Second Law establishes a quantitative relationship between force, mass, and acceleration. This law provides insights into how forces act upon objects, causing changes in their motion. It serves as a foundational principle for understanding dynamics and analyzing systems involving forces. Newton's Second Law finds practical applications across various fields, guiding engineering designs and enhancing our understanding of the physical world. It remains a cornerstone of classical mechanics and continues to shape our knowledge of the interplay between forces, mass, and motion.

— ◆ —

Newton's Third Law, Law of Action and Reaction

Newton's Third Law, also known as the Law of Action and Reaction, is a fundamental principle in physics that describes the relationship between the forces acting on interacting objects. This law states that for every action force exerted by one object on another, there is an equal and opposite reaction force exerted by the second object.

Definition of Newton's Third Law: Newton's Third Law states that when two objects interact, the forces they exert on each other are equal in magnitude but opposite in direction. It asserts that any force applied by one object will always result in an equal force, acting in the opposite direction, exerted by the other object. In simpler terms, it describes the balanced and reciprocal nature of forces between interacting objects.

Imagine you are sitting in a rowboat on a calm lake. When you push your hands against the water, you create an action force in one direction. The Law of Action and Reaction says that the water pushes back on your hands with an equal force in the opposite direction. This is the reaction force.

Again, this means that when you exert a force on an object, that object exerts an equal force back on you, but in the opposite direction. If you push a wall with a certain force, the wall pushes back on you with the same force.

This law is essential because it explains why objects stay still or move. When you walk, your feet push backward on the ground, and the ground pushes forward on your feet, propelling you forward. It's also why rockets can move in space—they push gas backward, and the gas pushes the rocket forward.

Example of Newton's Third Law in Real Life: An example of Newton's Third Law in real life is observed when we walk. As we take a step forward, our foot exerts a force backward against the ground. According to Newton's Third Law, the ground exerts an equal and opposite reaction force on our foot, propelling us forward. The action of pushing against the ground generates a reaction force that allows us to move in the desired direction.

Importance of Newton's Third Law: Newton's Third Law holds significant importance for several reasons:

1. Conservation of Momentum: Newton's Third Law is linked to the conservation of momentum. When two objects interact, the total momentum of the system remains constant. As one object exerts a force on the other, the forces cancel each other out, resulting in a net change of zero momentum for the system. This principle is crucial in understanding and analyzing the motion of objects and systems.

2. Balanced Forces and Equilibrium: Newton's Third Law highlights the balanced nature of forces in interactions. Forces always occur in pairs, and these pairs act simultaneously. This balance of forces contributes to the concept of equilibrium, where the net force on an object is zero. Understanding this law helps in identifying and analyzing equilibrium conditions in various systems, from static structures to dynamic systems.

3. Rocket Propulsion and Reaction Forces: Newton's Third Law is integral to rocket propulsion. A rocket exerts a powerful force downward by expelling exhaust gases at high velocities. As per Newton's Third Law, an equal and opposite reaction force propels the rocket upward. This law underlies the fundamental principle of action-reaction pairs in propulsion systems and is vital in space exploration and engineering.

4. Understanding Interactions and Collisions: Newton's Third Law provides insights into interactions between objects and their consequences during collisions. When two objects collide, the forces they exert on each other are equal and opposite. This understanding helps in analyzing and predict-

ing the outcomes of collisions, such as the conservation of energy and momentum.

Conclusion: Newton's Third Law describes the law of action and reaction. It asserts that for every action force exerted by one object on another, there is an equal and opposite reaction force exerted by the second object. This law highlights the balanced nature of forces in interactions and is integral to understanding momentum conservation, equilibrium, propulsion systems, and collision analysis. Newton's Third Law continues to shape our understanding of the dynamics of interacting objects and is a fundamental principle in classical physics.

17

FIRST LAW OF THERMODYNAMICS

The First Law of Thermodynamics, also known as the Law of Energy Conservation, is a fundamental principle in physics that states energy cannot be created or destroyed, but it can be converted from one form to another. This law provides a crucial foundation for understanding the behavior of energy in various systems and processes.

Definition of the First Law of Thermodynamics: The First Law of Thermodynamics states that energy is conserved within a closed system. It asserts that energy can neither be created nor destroyed; it can only be transferred or transformed from one form to another. In other words, the total energy of a system remains constant, and any changes in the system's internal energy are accounted for by energy transfer or conversion.

Imagine you have a bowl of hot soup. As you eat the soup and digest it, your body converts the energy stored in the food into different forms of energy, such as heat and the energy you use to move and think. The First Law of Thermodynamics tells us that the total amount of energy in the universe remains constant, even though it may change from one type of energy to another.

This law is crucial because it helps us understand the fundamental nature of energy and how it behaves in various processes, from simple everyday activities to complex scientific phenomena. It plays a central role in understanding how energy is used and transformed in everything around us, from the flow of heat in engines to the functioning of our bodies.

Credit to Rudolf Clausius: The First Law of Thermodynamics is a culmination of the works of several scientists, including Rudolf Clausius. Rudolf Clausius, a German physicist, introduced the concept of energy conservation in the context of thermodynamics in the mid-19th century. While the concept of energy conservation was already recognized, Clausius helped clarify and formalize it into a fundamental principle of physics.

Example of the First Law of Thermodynamics in Real Life: An example of the First Law of Thermodynamics in real life is the operation of an automobile engine. When fuel is burned within the engine, the chemical energy stored in the fuel is converted into thermal energy. This thermal energy is then partially transformed into mechanical energy to propel the vehicle forward. The First Law of Thermodynamics assures us that the total energy within the system, including the energy transferred and converted, remains constant.

Importance of the First Law of Thermodynamics: The First Law of Thermodynamics holds significant importance for several reasons:

1. Conservation of Energy: The First Law of Thermodynamics is rooted in the principle of energy conservation, one of the fundamental principles of physics. It establishes that energy is a conserved quantity, allowing us to analyze and

understand energy transfers and transformations in diverse systems. This law provides a foundation for studying heat, work, and other forms of energy in various scientific and engineering applications.

2. Energy Accounting: The First Law of Thermodynamics allows for the accounting of energy in a system. It enables scientists and engineers to quantify the energy inputs and outputs, track energy flow, and ensure that energy is conserved throughout processes. This accounting is crucial for optimizing energy usage, designing efficient systems, and assessing the overall energy balance in different applications.

3. Interplay of Heat and Work: The First Law of Thermodynamics reveals the interplay between heat and work in energy transfers. It allows us to understand how thermal energy can be converted into mechanical work, as well as how mechanical work can be converted into thermal energy. This understanding is vital in a wide range of fields, including power generation, heat engines, and energy conversion devices.

4. Basis for Thermodynamic Analysis: The First Law of Thermodynamics forms the basis for thermodynamic analysis, providing a framework for studying the behavior of systems and processes. It enables the development of thermodynamic models and equations, which allow us to predict and analyze energy transfers and transformations. These models find applications in fields such as engineering, chemistry, and environmental science.

Conclusion: The First Law of Thermodynamics represents the law of energy conservation within a closed system. It establishes that energy cannot be created or destroyed but can only be transferred or transformed. This law is vital for understanding energy flows, accounting for energy within systems, and analyzing energy transfers and transformations. The First Law of Thermodynamics continues to shape our understanding of the behavior of energy and plays a fundamental role in various scientific, engineering, and environmental applications.

18

LAWS OF CARTOON PHYSICS – NOT MEANT TO BE TAKEN SERIOUSLY

The Laws of Cartoon Physics are a set of whimsical principles that govern the fantastical world of animated cartoons. These laws define the boundaries and possibilities of the cartoon universe, allowing for extraordinary and often exaggerated phenomena that defy the laws of conventional physics.

Definition of the Laws of Cartoon Physics: The Laws of Cartoon Physics are a fictional set of principles that govern the physical world depicted in animated cartoons. They often involve humorous distortions of reality, allowing characters to bend or break the laws of traditional physics for comedic effect. These laws establish a unique and vibrant universe where characters can perform extraordinary feats, undergo elastic deformations, defy gravity, and more.

The Laws of Cartoon Physics are not real scientific principles but are designed to entertain and make us laugh with ideas such as:

Law of Gravity Defiance: In cartoons, characters can often defy gravity. They might float in mid-air, fall from great heights without getting hurt, or even walk on walls and ceilings.

Law of Elasticity: Cartoon characters can stretch and squash like rubber, allowing them to deform their bodies into all sorts of funny shapes.

Law of Inertia Ignorance: In cartoons, characters can suddenly stop or change direction without any regard for Newton's laws of motion.

Law of Acme Devices: Cartoon characters have access to all sorts of crazy and improbable gadgets and inventions that can solve any problem, no matter how absurd.

Law of Instant Appearance: Characters can magically appear out of thin air, disappear in a puff of smoke, or teleport to different places instantly.

Law of Exaggerated Physics: In cartoons, objects and characters can change size, grow or shrink at will, and perform incredible feats that defy the laws of the real world.

Creators of the Laws of Cartoon Physics: The Laws of Cartoon Physics do not have a specific individual credited with their creation. Rather, they have evolved over time through the collective creativity and imagination of cartoonists, animators, and storytellers. Icons such as Walt Disney, Tex Avery, Chuck Jones, and Hanna-Barbera have contributed to the establishment and development of these laws through their innovative and imaginative cartoons.

Example of the Laws of Cartoon Physics in Real Life: An example of the Laws of Cartoon Physics in real life can be observed when an individual slips on a banana peel. In cartoons, slipping on a banana peel often results in characters defying gravity, remaining suspended in mid-air for an extended period before finally succumbing to the inevitable fall. This comedic effect, inspired by the laws of cartoons, brings humor and entertainment to the situation.

Importance of the Laws of Cartoon Physics: The Laws of Cartoon Physics hold significance for several reasons:

1. Entertainment and Humor: The Laws of Cartoon Physics are essential for creating an entertaining and comedic experience for audiences. These laws allow for imaginative and exaggerated scenarios that defy the constraints of reality. They evoke laughter, surprise, and joy, bringing levity to storytelling and enhancing the appeal of animated cartoons.

2. Unleashing Imagination: The Laws of Cartoon Physics unleash the boundless power of imagination. By suspending the rules of conventional physics, these laws encourage viewers to embrace a world of limitless possibilities. They inspire creativity and allow for the exploration of alternative realities, expanding the horizons of storytelling and captivating the imaginations of both young and old.

3. Visual Expressions and Visual Gags: The Laws of Cartoon Physics provide a framework for visual expressions and gags. They enable animators to convey emotions, actions, and situations through exaggerated physical movements and transformations. These laws allow for visual comedy, slapstick humor, and memorable moments that have become iconic in the world of animation.

4. Suspension of Disbelief: The Laws of Cartoon Physics allow audiences to suspend their disbelief and immerse themselves in a fictional world where the impossible becomes possible.

They provide an escape from the constraints of reality, allowing viewers to experience moments of joy, wonder, and laughter. The suspension of disbelief is crucial for engaging audiences and creating a memorable and enjoyable entertainment experience.

Conclusion: The Laws of Cartoon Physics provide the rules and boundaries for the fantastical world of animated cartoons. They allow for imaginative and comedic scenarios that defy the laws of conventional physics, unleashing the power of imagination and entertaining audiences of all ages. These laws contribute to the magic and charm of animated cartoons, bringing laughter, wonder, and joy to those who embark on animated adventures. The Laws of Cartoon Physics continue to inspire creativity, captivate imaginations, and remind us of the sheer delight that can be found in the realm of animated storytelling.

PART 3

CHEMISTRY

CHEMISTRY

Chemistry is the branch of science that investigates the properties, structure, and transformations of matter. It explores how and why substances combine or separate to form other substances and how substances interact with energy. The field of chemistry is often divided into five main branches: organic, inorganic, analytical, physical, and biochemistry, each focusing on different aspects of matter and its transformations.

Chemistry is vitally important for numerous reasons. It plays a pivotal role in understanding the natural world around us, everything from why leaves change color to how our bodies use food. The principles of chemistry underpin all biological processes and are crucial to understanding global challenges such as climate change and pollution. Beyond the natural world, chemistry has profound implications in many industries, leading to the development of new materials, drugs, and energy sources. Its principles are integral to emerging fields like nanotechnology and environmental science. Furthermore, understanding chemistry can inform everyday decisions, like choosing cleaning products, understanding food labels, or assessing medical information. In essence, chemistry is a fundamental science that greatly enriches our knowledge and significantly impacts the quality and sustainability of our lives.

19

LAW OF CONSERVATION OF MASS

The Law of Conservation of Mass is a fundamental principle in chemistry and physics that states that mass cannot be created or destroyed in a chemical reaction or a physical process. It emphasizes the idea that the total mass of a closed system remains constant before and after any transformation or change.

Definition of the Law of Conservation of Mass: The Law of Conservation of Mass, also known as the Law of Conservation of Matter, states that in any closed system, the total mass before and after a chemical reaction or physical process remains constant. This law implies that matter cannot be created or destroyed but only transformed from one form to another. It provides the foundation for understanding the quantitative aspects of chemical reactions and the behavior of matter in various processes.

Imagine you have a sealed container with some water and ice inside. As you heat the container, the ice melts, and the water evaporates, eventually turning into steam. The Law of Conservation of Mass tells us that even though the form of the water is changing from solid ice to liquid water and then to gaseous steam, the total mass of the water remains the same throughout the process.

This law is essential because it helps scientists understand and predict chemical reactions and physical processes. It allows us to track the amount of matter in a system, in any form, solid, liquid, or gas, and ensures that mass is conserved in all natural processes. The Law of Conservation of Mass is a foundational principle that guides our understanding of matter and its behavior in various natural and man-made systems.

Credit to Antoine Lavoisier: Antoine Lavoisier, a French chemist, is credited with formulating the Law of Conservation of Mass in the late 18th century. Lavoisier conducted numerous experiments, particularly in combustion reactions, and observed that the total mass of substances involved in a reaction remained constant. He proposed the idea that matter is conserved during chemical reactions, establishing the concept of mass conservation in chemistry.

Example of the Law of Conservation of Mass in Real Life: An example of the Law of Conservation of Mass in real life is the burning of a piece of wood. When wood is burned, it undergoes a chemical reaction with oxygen from the air, resulting in the formation of carbon dioxide and water vapor. Despite the transformation of the wood, the total mass of the resulting products (carbon dioxide, water vapor, and any remaining ash) is equal to the initial mass of the wood. This example illustrates the conservation of mass during a chemical reaction.

Importance of the Law of Conservation of Mass: The Law of Conservation of Mass holds significant importance for several reasons:

> 1. Fundamental Principle of Chemistry: The Law of Conservation of Mass is a fundamental principle in chemistry. It provides a quantitative basis for understanding and analyz-

ing chemical reactions. By ensuring that mass is conserved during reactions, this law allows chemists to calculate the amount of reactants needed or products formed, predict the outcome of reactions, and understand the stoichiometry of chemical equations.

2. Basis for Stoichiometry: The Law of Conservation of Mass serves as the basis for stoichiometry, the branch of chemistry concerned with the quantitative relationships between reactants and products in chemical reactions. It allows scientists to determine the relative quantities of substances involved in a reaction, calculate reaction yields, and interpret the mass changes occurring during chemical transformations.

3. Environmental Impact and Resource Management: The Law of Conservation of Mass has implications for environmental science and resource management. It highlights the importance of waste management and the need to minimize the generation and disposal of materials. Understanding the law helps in designing processes that optimize resource utilization, reduce waste generation, and promote sustainable practices.

4. Conservation Laws in Physics: The Law of Conservation of Mass is connected to other conservation laws in physics, such as the conservation of energy and the conservation of momentum. It reflects the broader principle of the conservation of matter and energy in the universe. The law provides a bridge between the realms of chemistry and physics, emphasizing the interplay between matter and energy.

Conclusion: The Law of Conservation of Mass asserts that mass cannot be created or destroyed in a chemical reaction or physical process. It remains a foundational principle in chemistry, providing the basis for understanding the quantitative aspects of chemical reactions. This law ensures that matter is conserved and transformed rather than lost or gained during processes. This Law has far-reaching implications for stoichiometry, resource management, environmental science, and our understanding of the interconnections between matter and energy. It continues to be a vital principle that guides our exploration and understanding of the behavior of matter in various scientific and industrial applications.

20

LAW OF MULTIPLE PROPORTIONS

The Law of Multiple Proportions is a fundamental principle in chemistry that elucidates the relationship between the masses of elements in compounds. This law states that when two elements combine to form different compounds, the ratio of their masses will always be in small whole numbers.

Definition of the Law of Multiple Proportions: The Law of Multiple Proportions states that when two elements combine to form different compounds, the ratio of the masses of one element that combine with a fixed mass of the other element will always be in small whole numbers. In other words, elements can combine in different proportions to form distinct compounds, and the ratio of their masses will exhibit a simple numerical relationship.

Imagine you have two elements, A and B. They can join to create different compounds, but they will always combine in simple whole number ratios. For example, one atom of A might combine with one atom of B to form one compound, and another atom of A might combine with two atoms of B to form a different compound.

The Law of Multiple Proportions tells us that if two elements can form more than one compound, the ratio of their masses in these compounds will always be a simple whole number. For instance, if the ratio of element A to element B is 1:1 in one compound, it might be 1:2 in another compound, but it won't be something like 1:1.5.

This law is significant because it helps chemists predict and understand the different combinations that elements can form and the consistent patterns that occur in chemical reactions. It provides valuable insights into the relationships between elements and compounds, contributing to our understanding of the composition and behavior of matter.

Credit to John Dalton: John Dalton, the prominent English chemist, is credited with formulating the Law of Multiple Proportions in the early 19th century. Dalton extensively studied the composition of chemical compounds and proposed atomic theory, which incorporated the concept of multiple proportions. He observed that elements combined in fixed ratios by mass, and his work laid the groundwork for understanding the quantitative aspects of chemical composition.

Example of the Law of Multiple Proportions in Real Life: An example of the Law of Multiple Proportions in real life can be observed in the combination of carbon and oxygen to form carbon monoxide (CO) and carbon dioxide (CO_2). In carbon monoxide, the ratio of carbon to oxygen by mass is 12:16, simplifying to 3:4. In carbon dioxide, the ratio becomes 12:32, simplifying to 3:8. The Law of Multiple Proportions reveals that the ratio of carbon to oxygen in both compounds can be expressed in small whole numbers.

Importance of the Law of Multiple Proportions: The Law of Multiple Proportions holds significant importance for several reasons:

1. Insight into Chemical Composition: The Law of Multiple Proportions provides insights into the composition of chemical compounds. It establishes that elements can combine in different ratios to form distinct compounds, enabling scientists to determine the possible combinations and proportions of elements in compounds. This understanding is crucial for identifying and characterizing chemical substances.

2. Validation of Atomic Theory: The Law of Multiple Proportions supports and validates Dalton's atomic theory. By observing the fixed ratios of elements in different compounds, the law reinforces the concept that matter is composed of discrete and indivisible particles (atoms). It strengthens the notion that elements combine in specific ratios based on the fundamental properties of atoms.

3. Predictability of Chemical Reactions: The Law of Multiple Proportions facilitates predicting the outcome of chemical reactions. By understanding the ratios in which elements combine, chemists can predict the stoichiometry of chemical reactions and calculate the masses of reactants and products. This predictive power aids in the design and optimization of chemical processes.

4. Fundamental Principle in Stoichiometry: The Law of Multiple Proportions forms a fundamental principle in stoichiometry, the quantitative study of chemical reactions. It

provides the basis for calculating the relative amounts of reactants and products involved in a reaction, determining reaction yields, and interpreting the masses and ratios in chemical equations. It is an essential tool for chemists to analyze and quantify chemical transformations.

Conclusion: The Law of Multiple Proportion highlights the relationship between the masses of elements in compounds. It demonstrates that elements can combine in different proportions to form distinct compounds, with the ratios of their masses exhibiting simple numerical relationships. This law provides insights into the composition of chemical substances, validates the atomic theory, aids in predicting chemical reactions, and serves as a fundamental principle in stoichiometry. The Law of Multiple Proportions continues to shape our understanding of chemical composition and plays a crucial role in the quantitative analysis and study of chemical reactions.

21

LAW OF DEFINITE PROPORTIONS

The Law of Definite Proportions, also known as the Law of Constant Composition, is a fundamental principle in chemistry that states that a given chemical compound always contains the same elements in fixed and definite proportions by mass. This law highlights the consistent and unchanging nature of the composition of a compound, regardless of its source or method of preparation.

Definition of the Law of Definite Proportions: The Law of Definite Proportions states that a pure compound is composed of elements that are always present in fixed and definite proportions by mass. Regardless of the amount or source of the compound, the ratio of the masses of the constituent elements remains constant. This law implies that the composition of a compound is a characteristic property and does not vary.

Imagine you have a delicious chocolate chip cookie recipe. The Law of Definite Proportions tells us that no matter how many cookies you make, the proportion of the ingredients will always be the same. For example, if the recipe calls for 1 cup of flour, 1/2 cup of butter, and 1/2 cup of chocolate chips, every batch of cookies you bake will have these ingredients in the same fixed proportions.

In the same way, chemical compounds always have a specific arrangement of elements in consistent proportions. For example, water (H_2O) is always made up of two hydrogen atoms and one oxygen atom, and the mass ratio of hydrogen to oxygen will always be the same, no matter where the water comes from.

This law is crucial because it helps chemists identify and define chemical compounds with precision. It assures us that the composition of a compound is well-defined and constant, regardless of the source or the amount of the substance. The Law of Definite Proportions is a fundamental concept that underlies much of our understanding of chemical reactions and the behavior of matter in the natural world.

Credit to Joseph Proust: Joseph Proust, a French chemist, is credited with establishing the Law of Definite Proportions in the late 18th century. Proust conducted extensive experiments on the composition of compounds and observed that the elements in a given compound are always combined in fixed proportions by amount. His work challenged the prevailing notion that compounds could have variable compositions.

Example of the Law of Definite Proportions in Real Life: An example of the Law of Definite Proportions in real life is the compound water (H_2O). Water always consists of hydrogen and oxygen in a fixed mass ratio. For every one part by mass of hydrogen, there are eight parts by mass of oxygen. Whether obtained from different sources or prepared through various methods, water will always exhibit a 1:8 mass ratio of hydrogen to oxygen. This example exemplifies the constancy of composition dictated by the Law of Definite Proportions.

Importance of the Law of Definite Proportions: The Law of Definite Proportions holds significant importance for several reasons:

1. Consistency of Composition: The Law of Definite Proportions emphasizes the consistent and unchanging nature of the composition of compounds. It establishes that the ratio of elements in a compound remains fixed regardless of the compound's source, purity, or method of preparation. This knowledge is essential for identifying and characterizing compounds and ensuring their reproducibility.

2. Verification of Chemical Reactions: The Law of Definite Proportions enables chemists to verify the occurrence of chemical reactions and assess their completeness. By determining the masses of reactants and products involved in a reaction, scientists can compare the expected ratios based on the law with the actual ratios observed experimentally. This verification process ensures the accuracy and validity of chemical reactions.

3. Foundation for Stoichiometry: The Law of Definite Proportions forms a foundation for stoichiometry, the quantitative study of chemical reactions. It allows chemists to determine the relative amounts of reactants and products involved in a reaction, calculate reaction yields, and interpret the mass relationships in chemical equations. Stoichiometry is crucial for understanding and predicting the outcome of chemical reactions.

4. Basis for Chemical Formulas: The Law of Definite Proportions provides the basis for determining the chemical for-

mulas of compounds. By analyzing the masses and proportions of elements in a compound, chemists can establish the simplest and most consistent ratio of atoms. Chemical formulas convey crucial information about the composition and structure of compounds and are essential for communication in the field of chemistry.

Conclusion: The Law of Definite Proportions underscores the fixed and unchanging nature of the composition of compounds. It establishes that a given compound always consists of elements in fixed proportions by mass, regardless of its source or method of preparation. This law is instrumental in verifying chemical reactions, providing a foundation for stoichiometry, and determining chemical formulas. The Law of Definite Proportions plays a crucial role in understanding the consistency and reproducibility of chemical composition, allowing chemists to explore and analyze the building blocks of matter with precision and accuracy.

22

THEORIES OF VALENCE

The Theories of Valence are a set of conceptual frameworks in chemistry that aim to explain how atoms combine to form chemical compounds through the concept of valence. These theories provide insights into the nature of chemical bonding, the arrangement of electrons, and the formation of stable molecules.

Definition of the Theories of Valence: The Theories of Valence are a collection of models and concepts that attempt to describe how atoms interact and form chemical bonds to create compounds. These theories focus on the concept of valence, which represents the combining power of an atom or group of atoms to form chemical bonds. Valence is associated with the number of electrons an atom can gain, lose, or share to achieve a stable electron configuration.

Imagine atoms as friendly puzzle pieces, and chemical bonds as the connections that hold these pieces together to form a bigger puzzle—the molecule. Theories of Valence help us figure out how these puzzle pieces fit together and why certain atoms prefer to bond with specific partners.

One of the earliest theories, the "Valence Bond Theory," suggests that atoms share electrons to create chemical bonds. It's like each atom lending and borrowing puzzle pieces to create a stable and complete puzzle.

This theory helps chemists predict how atoms will behave when they come together to form molecules. They are like guiding principles that explain why some elements combine easily, while others prefer to stay on their own. By understanding the Theories of Valence, scientists can unlock the secrets of countless chemical reactions, build new materials, and create amazing substances that make our world work and look the way it does.

Credit to Gilbert N. Lewis and Linus Pauling: The Theories of Valence have evolved over time, with contributions from several scientists. Gilbert N. Lewis, an American chemist, proposed the concept of electron-pair bonds and developed the Lewis theory of valence in the early 20th century. Linus Pauling, another renowned American chemist, expanded on Lewis' work and formulated the concept of hybridization, which further enriched the understanding of chemical bonding.

Example of the Theories of Valence in Real Life: An example of the Theories of Valence in real life is the formation of methane (CH_4). According to the Lewis theory of valence, the central carbon atom in methane has four valence electrons. Each hydrogen atom contributes one electron, and carbon shares its remaining four electrons with the hydrogen atoms, forming four covalent bonds. This example demonstrates the concept of valence and how atoms combine to achieve a stable electron configuration through chemical bonding.

Importance of the Theories of Valence: The Theories of Valence hold significant importance for several reasons:

1. Understanding Chemical Bonding: The Theories of Valence provide a framework for understanding the nature of chemical bonding. They explain how atoms interact and form bonds by sharing, gaining, or losing electrons. These theories allow chemists to predict and analyze the types of bonds formed, such as covalent bonds, ionic bonds, and metallic bonds, and provide insights into the stability and reactivity of compounds.

2. Rationalizing Molecular Structures: The Theories of Valence help in rationalizing the structures of molecules. They allow chemists to determine the arrangement of atoms within a molecule, predict bond angles, and understand the spatial orientation of bonds. This knowledge aids in explaining the physical and chemical properties of compounds and guides the design of new molecules with desired properties.

3. Prediction of Chemical Reactivity: The Theories of Valence enable chemists to predict and understand the reactivity of compounds. By examining the valence electrons and their availability for bonding, these theories shed light on the likelihood of chemical reactions and the types of reactions that may occur. This predictive power aids in the synthesis of new compounds and the understanding of reaction mechanisms.

4. Development of Molecular Orbital Theory: The Theories of Valence have contributed to the development of more advanced theories, such as molecular orbital theory. These

theories explore the behavior of electrons in molecules using mathematical models, providing a deeper understanding of electronic structure, bonding, and spectroscopic properties. Molecular orbital theory builds upon the concepts of valence and enhances our understanding of complex molecules.

Conclusion: The Theories of Valence provide valuable insights into chemical bonding and the formation of compounds. These theories allow chemists to understand the combining power of atoms, predict the types of bonds formed, rationalize molecular structures, and predict chemical reactivity. They serve as a foundation for further advancements in understanding chemical bonding, such as molecular orbital theory. The Theories of Valence continue to play a crucial role in unraveling the mysteries of chemical compounds and shaping our understanding of the world of molecules.

23

THEORIES OF MOLECULAR ORBITALS

The Theories of Molecular Orbitals provide a powerful framework for understanding the behavior of electrons in molecules. These theories describe the formation and properties of molecular orbitals, which are regions in space where electrons are most likely to be found. By analyzing the distribution and energy levels of these orbitals, the theories of molecular orbitals enable chemists to comprehend the electronic structure, bonding, and spectroscopic properties of molecules.

Definition of the Theories of Molecular Orbitals: The Theories of Molecular Orbitals are conceptual frameworks in chemistry that describe the behavior of electrons in molecules. These theories consider the combination of atomic orbitals from individual atoms to form molecular orbitals. Molecular orbitals represent the distribution of electron density in a molecule and are associated with specific energy levels. Theories of Molecular Orbitals allow for the analysis of bonding, antibonding, and nonbonding interactions between electrons in a

Imagine atoms as performers on a stage, and the molecular orbitals as the different seats in the audience where the electrons can sit. In this theory, atoms join to create new "seats" for electrons to occupy. These seats, known as molecular orbitals, are like musical chairs where electrons move around, creating new patterns of energy and stability within the molecule.

One of the main theories, the "Molecular Orbital Theory," tells us that when atoms come together, their atomic orbitals (the individual seats for electrons) combine and interact to form new molecular orbitals. These molecular orbitals can be higher or lower in energy compared to the original atomic orbitals, and they determine how the electrons move and distribute themselves within the molecule.

By understanding the Theories of Molecular Orbitals, chemists can predict how molecules behave, how they interact with each other, and why some substances have unique properties. These theories are like the script that guides the chemistry of the world, allowing scientists to design new materials, understand chemical reactions, and explore the amazing world of atoms and molecules.

Credit to Friedrich Hund, Robert S. Mulliken, and John C. Slater: The development of the Theories of Molecular Orbitals involved contributions from several scientists. Friedrich Hund, a German physicist, introduced Hund's rules, which describe the distribution of electrons in molecular orbitals. Robert S. Mulliken, an American physicist and chemist, and John C. Slater, an American physicist, made significant advancements in the theory of molecular orbitals by applying quantum mechanics principles to explain electron behavior in molecules.

Example of the Theories of Molecular Orbitals in Real Life: An example of the Theories of Molecular Orbitals in real life is the bonding in the diatomic molecule oxygen (O2). According to the molecular orbital theory, two oxygen atoms combine, and their atomic orbitals overlap to form molecular orbitals. In this case, the combination of the 2p orbitals results in the formation of a sigma bonding orbital and a sigma* antibonding orbital. The electrons fill the lower-energy bonding orbital first, ensuring the stability of the molecule. This example demonstrates the application of molecular orbital theory in understanding the electronic structure and bonding in oxygen molecules.

Importance of the Theories of Molecular Orbitals: The Theories of Molecular Orbitals hold significant importance for several reasons:

1. Understanding Electronic Structure: The Theories of Molecular Orbitals provide a comprehensive understanding of the electronic structure of molecules. They allow chemists to analyze the distribution of electrons, their energy levels, and their interactions within a molecule. This understanding is crucial for predicting and explaining the spectroscopic properties, stability, and reactivity of molecules.

2. Describing Bonding and Antibonding Interactions: The Theories of Molecular Orbitals help in elucidating the nature of chemical bonding. By considering the combination of atomic orbitals, these theories explain the formation of bonding and antibonding molecular orbitals. They provide insights into the strength and stability of chemical bonds and guide the understanding of molecular structure and reactivity.

3. Prediction of Spectroscopic Properties: The Theories of Molecular Orbitals aid in predicting and interpreting the

spectroscopic properties of molecules. By analyzing the energy levels and transitions of electrons within molecular orbitals, these theories explain the absorption and emission of electromagnetic radiation. This predictive power enables the identification and characterization of compounds based on their spectroscopic fingerprints.

4. Advancements in Quantum Chemistry: The Theories of Molecular Orbitals have contributed to the advancement of quantum chemistry as a field. These theories incorporate quantum mechanics principles to describe electron behavior and interactions in molecules. They have paved the way for the development of computational methods and software tools used in molecular modeling, drug design, and material science.

Conclusion: The Theories of Molecular Orbitals provide a powerful framework for understanding the behavior of electrons in molecules. These theories allow chemists to analyze the electronic structure, bonding, and spectroscopic properties of molecules. By considering the combination of atomic orbitals, molecular orbital theory provides insights into the nature of chemical bonding and helps in predicting spectroscopic properties. The Theories of Molecular Orbitals have played a vital role in advancing our understanding of molecular behavior and have shaped the field of quantum chemistry. They continue to guide research and innovation in diverse areas such as material science, drug discovery, and molecular engineering.

24

THEORIES OF CHEMICAL BONDING

The Theories of Chemical Bonding encompass a range of conceptual frameworks that aim to explain the nature and formation of chemical bonds between atoms. These theories provide insights into the forces that hold atoms together in molecules, allowing chemists to understand the stability, structure, and reactivity of compounds.

Definition of the Theories of Chemical Bonding: The Theories of Chemical Bonding refer to a collection of models and concepts that seek to explain the nature and formation of chemical bonds. These theories aim to elucidate the attractive forces that hold atoms together, resulting in the formation of molecules. They encompass various approaches, such as Lewis theory, valence bond theory, and molecular orbital theory, each offering unique perspectives on chemical bonding.

Imagine atoms as friendly puzzle pieces, and chemical bonds as the connections that hold these pieces together to create a bigger puzzle—the molecule. The Theories of Chemical Bonding help us understand how these puzzle pieces fit together and why certain atoms prefer to bond with specific partners.

One of the main theories is the "Covalent Bonding Theory," which explains how atoms share electrons to form bonds. It's like each atom lending and borrowing puzzle pieces to create a stable and complete puzzle.

Another important theory is the "Ionic Bonding Theory," which describes how atoms transfer electrons to create oppositely charged ions that are held together by electrostatic forces. It's like atoms giving and taking puzzle pieces to create a strong attraction between them.

There's also the "Metallic Bonding Theory," which explains how atoms in metals share their electrons with each other, creating a sea of mobile electrons that hold the metal ions together like a group hug.

By understanding the Theories of Chemical Bonding, chemists can predict how atoms will behave when they come together to form molecules. These theories are like guiding principles that explain why some elements combine readily, while others prefer to stay on their own. They help us unlock the secrets of countless chemical reactions, build new materials, and create amazing substances that make our world work and look the way it does.

Credit to Gilbert N. Lewis, Linus Pauling, and others: The development of the Theories of Chemical Bonding involved contributions from multiple scientists. Gilbert N. Lewis and Linus Pauling, Robert Mulliken and Friedrich Hund, all mentioned previously made important advancements in the field of chemical bonding theory.

Example of the Theories of Chemical Bonding in Real Life: An example of the Theories of Chemical Bonding in real life is the formation of the water molecule (H_2O). According to the Lewis theory of chemical bonding, oxygen has six valence electrons and can share two of them with two hydrogen atoms to form covalent bonds. The Lewis structure of water shows two electron pairs shared between the oxygen and each hydrogen atom. This example illustrates how the Theories of Chemical Bonding explain the sharing of electron pairs between atoms to form stable compounds.

Importance of the Theories of Chemical Bonding: The Theories of Chemical Bonding hold significant importance for several reasons:

1. Understanding Molecular Stability: The Theories of Chemical Bonding provide insights into the forces that hold atoms together in molecules. They explain the stability of compounds and why certain arrangements of atoms are energetically favorable. This understanding is crucial for predicting and explaining the properties and behavior of molecules.

2. Rationalizing Molecular Structure: The Theories of Chemical Bonding help in rationalizing the structure of molecules. By considering the types of chemical bonds present, the geometry of atoms, and the distribution of electrons, these theories allow chemists to determine the arrangement of atoms within a molecule. This knowledge aids in explaining molecular properties and guiding the synthesis of new compounds.

3. Predicting Chemical Reactivity: The Theories of Chemical Bonding aid in predicting and understanding the reactivity of compounds. By considering the strength and nature of chemical bonds, these theories provide insights into the likelihood of chemical reactions and the types of reactions that may occur. This predictive power is invaluable for designing

chemical reactions and understanding reaction mechanisms.

4. Guiding Material Design: The Theories of Chemical Bonding are essential in the design and development of new materials. By understanding how atoms bond and interact, chemists can manipulate the properties of materials by controlling their molecular structure. This knowledge is instrumental in fields such as material science, nanotechnology, and catalysis.

Conclusion: The Theories of Chemical Bonding offer valuable insights into the nature and formation of chemical bonds. These theories help in understanding molecular stability, rationalizing molecular structure, predicting chemical reactivity, and guiding material design. The Theories of Chemical Bonding are fundamental to the field of chemistry, providing a framework for understanding the forces that bind atoms together and allowing scientists to unravel the intricacies of chemical bonding. Their importance extends across various scientific disciplines and applications, making them indispensable tools in the pursuit of knowledge and innovation.

PART 4

BIOLOGY

BIOLOGY

*B*iology is the scientific study of life and living organisms. This broad field encompasses many specialized disciplines that cover various aspects of the complexity of life, from the smallest biological molecules to the interaction of millions of species in an ecosystem. These disciplines include genetics, where we study heredity and variation; microbiology, where we investigate microscopic organisms; ecology, which focuses on how organisms interact with each other and their environment; and many more. The field also explores vital processes such as growth, reproduction, metabolism, and evolution.

Biology holds profound importance for many reasons. It provides the foundational understanding of the human body, which informs medical advancements and healthcare. Without the study of biology, we wouldn't have the treatments and medicines we use to combat diseases. Additionally, understanding biological principles helps address global challenges like climate change, deforestation, and species extinction. It guides us in preserving biodiversity and maintaining ecological balance. Biology also plays a crucial role in the biotechnology industry, leading to developments in areas like genetically modified organisms (GMOs), biofuels, and cloning. Moreover, biology helps us understand the process of evolution and the interconnections in the web of life, which adds depth to our understanding of our own existence and role in the world. In *essence, the study of biology impacts virtually every aspect of our lives and holds the keys to many of the challenges we face today.*

25

CELL THEORY

Cell Theory is a foundational concept in biology that describes the fundamental structural and functional unit of all living organisms—the cell. This theory proposes that all living organisms are composed of cells, that cells are the basic units of structure and function, and that cells arise only from pre-existing cells.

Definition of Cell Theory: Cell Theory is a scientific principle that states three fundamental propositions: (1) all living organisms are made up of one or more cells, (2) cells are the basic units of structure and function in living organisms, and (3) cells can only arise from pre-existing cells. This theory emphasizes the central role of cells in the organization and functioning of living systems.

Imagine you have a puzzle, and each piece of the puzzle represents a living thing. The Cell Theory tells us that every living organism, from tiny bacteria to giant whales, is made up of small units called cells. These cells are like individual puzzle pieces that come together to create the whole picture of life.

In simpler terms, everything that is alive is made up of cells, and cells are the essential building blocks of life. Each cell has specific functions, like how your skin cells protect your body or how nerve cells help you feel and think. Additionally, cells can make copies of themselves to create new cells and keep living things growing and functioning.

The Cell Theory is crucial because it helps us understand the remarkable similarities and fundamental principles that govern life across different living organisms. It forms the foundation of modern biology and guides our understanding of how living things work, evolve, and interact with their environment.

Credit to Matthias Schleiden and Theodor Schwann: Cell Theory emerged through the contributions of multiple scientists. In the early 19th century, the German botanist Matthias Schleiden and the German physiologist Theodor Schwann independently proposed that plants and animals are composed of cells. Collectively, their work laid the foundation for Cell Theory.

Example of Cell Theory in Real Life: An example of Cell Theory in real life is the study of tissue samples. When examining a tissue sample, such as a piece of skin, cells are observed under a microscope. The tissue is composed of numerous cells, each performing specific functions within the skin, such as providing protection, sensing stimuli, or synthesizing certain substances. The presence of individual cells within the tissue confirms the principles of Cell Theory, highlighting their role as the building blocks of complex organisms.

Importance of Cell Theory: Cell Theory holds significant importance for several reasons:

1. Fundamental Concept of Biology: Cell Theory is a fundamental concept in biology, providing the basis for understanding the organization and functioning of living or-

ganisms. By recognizing cells as the basic units of life, the theory establishes a framework for studying the diversity and complexity of living systems. It serves as a unifying principle, allowing scientists to explore the commonalities among different organisms and their shared cellular characteristics.

2. Explaining Life Processes: Cell Theory helps explain various life processes. By recognizing that cells are the sites of essential functions such as metabolism, growth, reproduction, and response to stimuli, the theory provides insights into how living organisms function. It enables scientists to investigate the intricate processes occurring within cells and understand how they contribute to the overall functioning of an organism.

3. Advancement of Medical Science: Cell Theory has revolutionized medical science. By recognizing that diseases originate from abnormalities in cells, the theory has paved the way for the understanding and treatment of various diseases. It has enabled the development of diagnostic techniques, such as analyzing cells for signs of cancer, infections, or genetic disorders. Moreover, the theory has contributed to advancements in regenerative medicine and tissue engineering.

4. Evolutionary Perspective: Cell Theory offers an evolutionary perspective on the origin and diversity of life. By acknowledging that all cells arise from pre-existing cells, the theory aligns with the principles of evolutionary biology. It helps explain how life has evolved over billions of years through the accumulation of genetic variations within cells. Cell Theory supports the concept that all organisms share a common

ancestry and have evolved from a common cellular origin.

Conclusion: Cell Theory represents a cornerstone in the field of biology. This theory defines cells as the fundamental units of life, highlights their role in the organization and functioning of living organisms, and emphasizes their continuity through cell division. Cell Theory is integral to our understanding of the complexity and unity of life, explaining the diverse processes occurring within cells and providing a framework for studying the interconnections between cells and larger biological systems. It has shaped our knowledge of biology and continues to drive scientific exploration and medical advancements, unraveling the mysteries of life at its most basic level—the cell.

26

THEORY OF EVOLUTION

The Theory of Evolution is a fundamental concept in biology that explains the diversity of life on Earth and the processes by which species have evolved and continue to change over time. It proposes that all living organisms share a common ancestry and have gradually developed through natural selection, genetic variation, and the adaptation to changing environments.

Definition of the Theory of Evolution: The Theory of Evolution suggests that all species on Earth have descended from a common ancestor through a process of gradual change over vast periods of time. It implies that the diversity of life arises through the mechanisms of natural selection, genetic mutation, and genetic drift. Evolutionary theory explains how species adapt and change in response to their environment, leading to the formation of new species over time.

Imagine you have a family tree, but instead of just your immediate family, it includes all your ancestors going back hundreds of generations. The Theory of Evolution tells us that all living things have such a family tree too, connecting them to their ancient ancestors. This tree shows how species have evolved, or gradually changed, over millions of years to adapt to their environments and survive.

The main idea behind the Theory of Evolution is natural selection. It's like a game of "survival of the fittest." Living things that have traits better suited for their environment are more likely to survive, reproduce, and pass those helpful traits to their offspring. Over time, these helpful traits become more common in the population, leading to changes and adaptations in species.

For example, imagine a group of birds with different beak sizes. If they live on an island with large seeds, birds with bigger, stronger beaks can crack the seeds open more easily, making them more likely to find food and survive. Through generations, birds with larger beaks will become more common because they have a better chance of passing their beak size to their offspring.

The Theory of Evolution helps us understand the incredible diversity of life on Earth and how species have evolved and adapted to different environments over millions of years. It is one of the most important and well-supported theories in biology and has transformed our understanding of the natural world and our place in it.

Credit to Charles Darwin: The Theory of Evolution is primarily credited to Charles Darwin, an English naturalist and biologist, who published his seminal work "On the Origin of Species" in 1859. Darwin's extensive observations and meticulous documentation during his travels, coupled with his insight into natural selection, provided the foundation for the Theory of Evolution.

Example of the Theory of Evolution in Real Life: An example of the Theory of Evolution in real life is the evolution of the songbirds, the Galápagos finches. During his voyage on the ship the HMS Beagle, Darwin observed that finches on different Galápagos Islands displayed unique variations in beak size, shape, and functionality. He hypothesized that these variations were adaptations to the specific food sources available on each island. Later research confirmed that these finches had diverged from a common ancestor and undergone adaptive changes, leading to the evolution of distinct species with specialized beak structures for specific diets. This example exemplifies how the Theory of Evolution explains the process of natural selection and adaptation, leading to the formation of new species over time.

Importance of the Theory of Evolution: The Theory of Evolution holds significant importance for several reasons:

1. Understanding Biological Diversity: The Theory of Evolution provides a unifying framework for understanding the remarkable diversity of life on Earth. It explains how the branching patterns of evolution have resulted in the vast array of species and the interconnectedness of all living organisms. By studying evolutionary processes, scientists gain insights into the origins, relationships, and adaptations of species, contributing to our understanding of biodiversity.

2. Illuminating Anatomical and Genetic Similarities: The Theory of Evolution explains the anatomical and genetic similarities observed among different species. It reveals that these similarities are the result of shared ancestry and common descent. By studying homologous structures and DNA sequences, scientists can trace the evolutionary relationships between organisms, helping unravel the complex tree of life and providing insights into the fundamental biological principles that shape living systems.

3. Explaining Adaptation and Survival: The Theory of Evolution provides a framework for understanding how species adapt and survive in changing environments. Through the process of natural selection, organisms with traits that are advantageous for their environment are more likely to survive and reproduce, passing those favorable traits to future generations. This understanding is crucial for comprehending the dynamic interplay between organisms and their surroundings, guiding conservation efforts and the preservation of biodiversity.

4. Guiding Medical Research and Applications: The Theory of Evolution has significant implications for medical research and applications. It helps explain the emergence of antibiotic resistance in bacteria, the evolution of viruses, and the genetic basis of diseases. By understanding the evolutionary history of pathogens, scientists can develop effective strategies for disease prevention, treatment, and the development of vaccines.

Conclusion: The Theory of Evolution provides a comprehensive and elegant explanation for the diversity and interconnectedness of life on Earth. It explains the mechanisms of natural selection, genetic variation, and adaptation that have shaped species over millions of years. This theory is fundamental to understanding the unity of all living organisms, explaining anatomical and genetic similarities, and unraveling the complex relationships among species. The Theory of Evolution continues to be a cornerstone of modern biology, influencing scientific research, conservation efforts, medical advancements, and our understanding of the intricate web of life.

27

THEORY OF GENETICS

The Theory of Genetics is a fundamental concept in biology that explains how traits are passed from one generation to the next through the transmission of genetic material. This theory encompasses the study of genes, heredity, and the mechanisms underlying the inheritance of traits. It provides insights into the genetic basis of diversity and the interplay between genes and the environment.

Definition of the Theory of Genetics: The Theory of Genetics is a scientific framework that describes the principles and mechanisms of heredity. It includes the study of genes, which are segments of DNA that contain instructions for the development and functioning of organisms. The theory explains how genes are inherited, how they interact to produce traits, and how genetic variations contribute to the diversity observed within and among species.

In other words, the Theory of Genetics is a scientific explanation that explores how traits are passed down from one generation to another in living organisms.

Imagine you have a puzzle representing your traits, like your eye color, hair color, or height. The Theory of Genetics tells us that these traits are inherited from your parents. Just like you get pieces of a puzzle from both your mom and dad, you get genes from both of them that carry instructions for your traits.

Genes are like tiny instructions or recipes inside your body's cells. They determine how you look and function. When a baby is born, it inherits a combination of genes from its parents, creating a unique blend of traits.

The Theory of Genetics is crucial because it helps us understand the mechanisms behind heredity and the passing on of traits from one generation to the next. It also explains how genetic variations can lead to differences among individuals and how genetic factors play a role in health, disease, and other aspects of life.

By studying genetics, scientists can learn more about how living things are related and how characteristics are inherited, leading to advancements in medical research, agriculture, and our overall understanding of life and its incredible diversity.

Credit to Gregor Mendel: The Theory of Genetics owes much of its development to the pioneering work of Gregor Mendel, an Austrian monk and botanist, in the mid-19th century. Mendel conducted groundbreaking experiments with pea plants, carefully observing and quantifying the transmission of traits from one generation to the next. His work laid the foundation for the understanding of genetic inheritance, establishing principles such as segregation and independent assortment.

Example of the Theory of Genetics in Real Life: An example of the Theory of Genetics in real life is the inheritance of eye color in humans. Eye color is determined by multiple genes, and the specific combination of these genes contributes to the observed variations in eye color. According to the Theory of Genetics, each parent contributes one copy of their genes to their offspring. Through the process of genetic recombination, the offspring inherits a unique combination of genes, resulting in the expression of a specific eye color. This example demonstrates how the Theory of Genetics explains the transmission and variation of traits in humans.

Importance of the Theory of Genetics: The Theory of Genetics holds significant importance for several reasons:

1. Understanding Inheritance: The Theory of Genetics provides a comprehensive understanding of how traits are inherited and passed from one generation to the next. It explains the principles of dominance, segregation, and independent assortment, elucidating the patterns and mechanisms of inheritance. This knowledge allows scientists to predict and understand the likelihood of certain traits appearing in offspring and trace the transmission of genetic disorders.

2. Unraveling the Genetic Basis of Traits: The Theory of Genetics enables scientists to explore the genetic basis of traits and understand how specific genes contribute to the observed variations in organisms. By studying the interactions between genes and their expression, researchers can uncover the genetic mechanisms underlying physical characteristics, behaviors, and susceptibility to diseases. This knowledge aids in the diagnosis, treatment, and prevention of genetic disorders.

3. Advancements in Biotechnology and Genetic Engineering: The Theory of Genetics has led to significant advancements in biotechnology and genetic engineering. It has provided the foundation for techniques such as gene sequencing, genetic modification, and gene therapy. These technologies allow scientists to manipulate and modify genetic material, leading to advancements in agriculture, medicine, and the development of novel therapies.

4. Conservation and Biodiversity Studies: The Theory of Genetics plays a crucial role in conservation efforts and the preservation of biodiversity. By studying genetic diversity within populations, scientists can assess the health, viability, and evolutionary potential of species. Genetic information helps identify distinct populations, track migration patterns, and guide conservation strategies to protect endangered species and maintain ecosystem stability.

Conclusion: The Theory of Genetics is a fundamental concept that reveals the principles and mechanisms of heredity. It explains how genes are transmitted, how they interact to produce traits, and how genetic variations contribute to the diversity observed in living organisms. The Theory of Genetics enables the understanding of inheritance patterns, unravels the genetic basis of traits, drives advancements in biotechnology, and aids in conservation efforts. This theory forms the cornerstone of modern genetics and plays a vital role in various fields, from medicine and agriculture to biodiversity conservation and our understanding of life's complexity.

28

THEORY OF ECOLOGY

The Theory of Ecology is a foundational concept in biology that focuses on the study of the relationships between organisms and their environment. It explores the interdependencies and interactions within ecosystems, emphasizing the flow of energy, the cycling of nutrients, and the conservation of biodiversity.

Definition of the Theory of Ecology: The Theory of Ecology encompasses the study of the relationships between organisms and their environment, examining the interplay between living organisms, their habitats, and the physical and biological factors that shape their interactions. It investigates the distribution, abundance, and behavior of organisms, the flow of energy through ecosystems, and the cycling of nutrients, emphasizing the interconnectedness of all living organisms within their ecological communities.

Imagine you have a vast orchestra, and each musician represents a different living thing, such as plants, animals, and microorganisms. The Theory of Ecology tells us that all these musicians play different instruments and notes, but together, they create a beautiful symphony that harmoniously resonates throughout the natural world. Each living thing in the orchestra has a crucial role, contributing to the balance and richness of the ecosystem. Just like how the orchestra's pieces come together to create a masterpiece, the interactions between living organisms form a complex and interconnected web of life that sustains our planet.

Ecology explores how living things interact with one another, such as how animals hunt for food or how plants compete for sunlight and space. It also examines how living things interact with non-living things in their environment, like how animals adapt to changes in temperature or how plants respond to rainfall.

This theory helps us understand the balance and harmony that exists in nature. Just like how music notes fit together to create a beautiful song, the interactions between living organisms and their environment create a complex and dynamic ecosystem.

By studying ecology, scientists can gain insights into how ecosystems function, how they respond to changes, and how we, as humans, can live in cooperation with the natural world. It is a fundamental field of study that helps us appreciate the intricate web of life on Earth and the importance of protecting and preserving our environment for future generations.

Credit to Ernst Haeckel, Eugene Odum, and G. Evelyn Hutchinson: The Theory of Ecology has evolved over time, with contributions from various scientists. Ernst Haeckel, a German biologist, coined the term "ecology" in the late 19th century and laid the groundwork for the study of the interrelationships between organisms and their environment. Eugene Odum, an American ecologist, further developed the theory and is often referred to as the "father of modern ecology." G. Evelyn Hutchinson, another influential ecologist, expanded ecological theory by emphasizing the complex dynamics and interactions within ecosystems.

Example of the Theory of Ecology in Real Life: An example of the Theory of Ecology in real life is the study of predator-prey interactions in a grassland ecosystem. In this ecosystem, the presence of predators, such as lions or cheetahs, influences the population dynamics of their prey, such as zebras or gazelles. The Theory of Ecology explains how changes in predator populations can affect the abundance and behavior of their prey. Predators control herbivore populations, preventing overgrazing and maintaining the balance within the ecosystem. This example demonstrates how the Theory of Ecology explains the interdependencies and interactions between organisms and their environment.

Importance of the Theory of Ecology: The Theory of Ecology holds significant importance for several reasons:

1. Understanding Ecosystem Dynamics: The Theory of Ecology provides a framework for understanding the dynamics of ecosystems. It explains the flow of energy through food chains and food webs, the cycling of nutrients, and the complex interactions between organisms within their environment. By studying ecological processes, scientists gain insights into the stability, resilience, and functioning of ecosystems, contributing to our understanding of the natural world.

2. Guiding Conservation and Resource Management: The Theory of Ecology plays a crucial role in guiding conservation efforts and the sustainable management of natural

resources. By understanding the interdependencies between organisms and their habitats, scientists can develop effective strategies for conserving biodiversity, restoring ecosystems, and mitigating the impacts of human activities. Ecology helps identify keystone species, evaluate the consequences of habitat destruction, and assess the effects of climate change on ecosystems.

3. Exploring Species Interactions and Adaptations: The Theory of Ecology allows for the exploration of species interactions and adaptations. It helps unravel the mechanisms of coexistence, competition, predation, and mutualism among organisms. By studying how species respond to environmental changes, scientists gain insights into their adaptations, reproductive strategies, and evolutionary dynamics. This knowledge aids in understanding the complexity of ecological communities and the evolutionary processes that shape biodiversity.

4. Addressing Environmental Challenges: The Theory of Ecology is crucial for addressing global environmental challenges. By understanding the impacts of human activities on ecosystems, such as pollution, habitat destruction, and climate change, scientists can develop sustainable practices and policies. Ecology provides the knowledge and tools to assess environmental risks, restore degraded habitats, and promote the conservation of biodiversity, ensuring the long-term health of our planet.

Conclusion: The Theory of Ecology offers a comprehensive understanding of the relationships between organisms and their environment. It explores the interdependencies, energy flow, and nutrient cycling within ecosystems, guiding our understanding of the natural world and the conservation of biodiversity. The Theory of Ecology provides insights into ecosystem dynamics, informs conservation and resource management, elucidates species interactions and adaptations, and addresses pressing environmental challenges. It serves as a crucial tool for exploring the intricate web of life and promoting a sustainable coexistence with our planet's ecosystems.

29

THEORY OF HOMEOSTASIS

The Theory of Homeostasis is a fundamental concept in biology that describes the dynamic process by which living organisms maintain internal stability in the face of changing external conditions. It encompasses the regulatory mechanisms and feedback loops that ensure the balance and stability of key physiological variables.

Definition of the Theory of Homeostasis: The Theory of Homeostasis states that living organisms possess internal regulatory mechanisms that work to maintain a stable and balanced internal environment, despite changes in the external surroundings. It involves a continuous process of monitoring, adjusting, and responding to internal and external stimuli, aiming to keep physical variables within a narrow range necessary for optimal functioning.

In other words, the Theory of Homeostasis explains how living organisms maintain a stable and balanced internal environment, even when faced with changes in the external surroundings.

Imagine you have a thermostat in your house. When the temperature inside gets too hot, the thermostat turns on the air conditioning to cool the room down. If it gets too cold, the thermostat switches on the heater to warm it up. The Theory of Homeostasis works similarly in living things.

Living organisms have internal control systems that monitor different aspects of their bodies, like temperature, water levels, and blood sugar. When any of these factors deviate from their normal range, the internal control systems kick in to bring them back to balance. It's like the body's own built-in thermostat that keeps things in check.

For example, when you exercise and your body temperature rises, your internal control system responds by making you sweat, which cools you down. If your blood sugar gets too high after eating, your body releases insulin to lower it back to normal levels.

The Theory of Homeostasis is essential because it ensures that living things can function optimally and stay healthy, even in changing environments. It's like having a self-regulating system that keeps everything in harmony, allowing organisms to adapt and survive in their surroundings.

Credit to Claude Bernard and Walter Cannon: The Theory of Homeostasis has roots in the work of several scientists, but two prominent figures stand out. Claude Bernard, a French physiologist in the 19th century, first proposed the concept of the internal environment and its constancy as a crucial aspect of maintaining health. Walter Cannon, an American physiologist, further developed the theory and coined the term "homeostasis" to describe the dynamic equilibrium maintained by living organisms.

Example of the Theory of Homeostasis in Real Life: An example of the Theory of Homeostasis in real life is the regulation of body temperature in humans. When external temperatures rise, the body initiates cooling mechanisms such as sweating and vasodilation (widening of blood vessels), allowing excess heat to dissipate. On the other hand, when external temperatures drop, the body activates mechanisms such as shivering and vasoconstriction (narrowing of blood vessels) to conserve heat and maintain internal temperature. These physiological responses illustrate the Theory of Homeostasis in action, as the body constantly adjusts to maintain its internal temperature within a narrow range.

Importance of the Theory of Homeostasis: The Theory of Homeostasis is of great importance for several reasons:

1. Optimal Physiological Functioning: Homeostasis ensures the optimal functioning of living organisms by maintaining the internal environment required for cellular activities. It regulates variables such as temperature, pH, electrolyte balance, and nutrient levels within narrow ranges that support proper enzymatic reactions and cellular processes. This stability is essential for the survival and well-being of organisms.

2. Adaptation to Changing Conditions: The Theory of Homeostasis enables organisms to adapt and respond to changing external conditions. It allows living organisms to cope with fluctuations in their surroundings while maintaining stability internally. This adaptability is particularly crucial for survival in environments characterized by seasonal variations, temperature extremes, or limited resources.

3. Health and Disease Prevention: Homeostasis plays a vital role in maintaining health and preventing disease. Imbalances or disruptions in homeostatic mechanisms can lead to

pathological conditions. For instance, diabetes results from a disruption in the regulation of blood glucose levels. Understanding homeostasis helps in identifying the underlying causes of diseases and developing strategies to restore balance and promote well-being.

4. Biological Research and Medicine: The Theory of Homeostasis provides a framework for biological research and medical interventions. It allows scientists and healthcare professionals to study and understand the mechanisms by which organisms maintain stability and respond to disruptions. This knowledge guides the development of diagnostic tests, therapies, and treatments aimed at restoring homeostatic balance in diseased states.

Conclusion: The Theory of Homeostasis describes the dynamic process by which living organisms maintain stability and balance in their internal environment. It ensures optimal physiological functioning, enables adaptation to changing conditions, promotes health, and guides medical interventions. The Theory of Homeostasis is a cornerstone of biological understanding, emphasizing the delicate equilibrium necessary for life. It has broad applications in various fields, from basic research to clinical medicine, and continues to shape our knowledge of how living organisms regulate their internal environment to thrive in diverse and ever-changing circumstances.

30

LAW OF THE MINIMUM

The Law of the Minimum is a principle that explains the growth and maturation of plants based on the availability of essential nutrients. This law states that plant growth is limited by the nutrient that is in the shortest supply relative to the plant's needs.

Definition of the Law of the Minimum: The Law of the Minimum posits that the growth and development of plants are determined by the availability of essential nutrients. According to this principle, if any necessary nutrient is deficient or limited in supply, it becomes the limiting factor that hinders optimal growth, regardless of the abundance of other nutrients. In other words, plant growth is restricted by the nutrient that is most scarce in relation to the plant's requirements.

Imagine a plant needing water, sunlight, and nutrients from the soil to grow. If it has plenty of sunlight and nutrients but lacks water, its growth will be limited by the lack of water, making it the "limiting factor." Even if the plant has an abundance of the other resources, its overall growth will be constrained by the minimum amount of water it needs.

This concept is crucial in agriculture and gardening because it reminds us that if any essential resource is insufficient, it can hinder the growth of plants. By identifying and providing the limiting factor, we can optimize plant growth and ensure healthy and thriving vegetation.

Credit to Justus von Liebig: The Law of the Minimum is credited to Justus von Liebig, a renowned German chemist and agronomist who is considered one of the founders of modern agricultural chemistry. Von Liebig developed this law in the mid-19th century based on his extensive research on plant nutrition and the relationships between soil fertility and crop production. His work significantly advanced the understanding of plant physiology and laid the groundwork for modern agricultural practices.

Example of the Law of the Minimum in Real Life: An example of the Law of the Minimum can be observed in crop production. Suppose a particular crop requires nitrogen, phosphorus, and potassium (NPK) for optimal growth. If the soil is deficient in potassium, even if there is an abundance of nitrogen and phosphorus, the plant's growth will be limited by the insufficient potassium supply. The plant will not be able to achieve its maximum potential growth until the potassium deficiency is addressed. This example illustrates how the availability of nutrients, specifically the limiting factor, determines plant growth and crop productivity.

Importance of the Law of the Minimum: The Law of the Minimum holds significant importance in the field of agriculture and plant nutrition:

> 1. Maximizing Crop Yield: Understanding the Law of the Minimum allows farmers and agronomists to optimize crop

production by identifying and addressing nutrient deficiencies. By ensuring that all essential nutrients are provided in sufficient quantities, farmers can maximize crop yields and improve overall agricultural productivity.

2. Nutrient Management: The law highlights the importance of balanced nutrient management in agriculture. It emphasizes the need to consider the specific nutrient requirements of different crops and the role of soil fertility in providing essential nutrients. Proper nutrient management helps prevent deficiencies and promotes healthy plant growth.

3. Soil Fertility Assessment: The Law of the Minimum underscores the significance of soil fertility assessments. By analyzing soil samples and identifying nutrient deficiencies, farmers can tailor their fertilizer application strategies to address specific nutrient limitations and maintain optimal soil fertility.

4. Sustainable Agriculture: The law promotes sustainable agricultural practices by encouraging the efficient use of nutrients. By understanding the role of limiting factors and nutrient interactions, farmers can adopt precision farming techniques, such as targeted fertilization and soil amendments, to minimize nutrient waste and environmental impacts.

5. Research and Innovation: The Law of the Minimum continues to inspire research and innovation in plant nutrition. Scientists are constantly exploring new approaches, such as nutrient management technologies, genetic improvements, and precision agriculture tools, to enhance nutrient avail-

ability and plant utilization, ultimately contributing to sustainable and productive agriculture.

Conclusion: The Law of the Minimum, formulated by Justus von Liebig, emphasizes the critical role of essential nutrients in plant growth and development. This principle highlights that plant growth is limited by the nutrient that is in the shortest supply relative to the plant's requirements. Understanding the Law of the Minimum is crucial for optimizing crop production, managing soil fertility, and promoting sustainable agricultural practices. By recognizing and addressing nutrient deficiencies, farmers can enhance yields, improve food security, and contribute to sustainable agriculture, ensuring the efficient utilization of resources while minimizing environmental impacts.

31

THEORY OF GROWTH AND DEVELOPMENT

The Theory of Growth and Development is a fundamental concept in biology that explores the sequential and progressive changes that occur in living organisms throughout their lifespan. It encompasses the physical, cognitive, and emotional transformations from conception to maturity.

Definition of the Theory of Growth and Development: The Theory of Growth and Development encompasses the study of the systematic and predictable changes that occur in living organisms from the moment of conception to maturity. It examines the physical, cognitive, social, and emotional aspects of development, highlighting the patterns, stages, and influences that shape an individual's growth and maturation.

Imagine you have a tiny plant seed. As you plant it in the soil and provide it with water and sunlight, it begins to sprout and grow into a small seedling. Over time, the seedling develops into a full-grown plant with leaves, stems, and flowers. This process of change and progression from a seed to a mature plant is an example of growth and development.

In the same way, animals, including humans, also go through various stages of growth and development. For instance, a baby starts as a tiny, helpless newborn and gradually grows into a toddler, child, teenager, and eventually an adult. Each stage brings different physical and behavioral changes, like learning to walk, talk, and interact with others.

The Theory of Growth and Development helps us understand the natural progression and changes that occur in living organisms as they mature and adapt to their environment. It's like watching a time-lapse video of a plant or an animal transforming and evolving throughout its life. By studying this theory, scientists can gain insights into the patterns and processes of life's journey, leading to a deeper understanding of the complexity and beauty of living organisms.

Credit to Jean Piaget and Erik Erikson: The Theory of Growth and Development has been shaped by the contributions of several researchers, with Jean Piaget and Erik Erikson being notable figures in this field. Jean Piaget, a Swiss psychologist, developed a comprehensive theory of cognitive development, emphasizing the progressive and qualitative shifts in thinking abilities from infancy to adulthood. Erik Erikson, a German-American psychologist, proposed a theory of psychosocial development that emphasizes the importance of social interactions and psychological conflicts in shaping individual development across the lifespan.

Example of the Theory of Growth and Development in Real Life: An example of the Theory of Growth and Development in real life is the development of language skills in children. According to Piaget's theory, children progress through different stages of cognitive development, including the preoperational stage. During this stage, typically occurring between the ages of 2 and 7, children acquire language skills and develop the ability to represent objects and events symbolically. This example illustrates how the Theory of Growth and Development helps explain the sequential acquisition of language skills in children and provides insights into the cognitive processes involved in their linguistic development.

Importance of the Theory of Growth and Development: The Theory of Growth and Development holds significant importance for several reasons:

1. Understanding Human Development: The Theory of Growth and Development provides a comprehensive understanding of the systematic changes that occur in individuals throughout their lifespan. It sheds light on the physical, cognitive, social, and emotional aspects of human development. This knowledge enables researchers, educators, and caregivers to better understand and support individuals at different stages of life, facilitating their overall well-being and growth.

2. Informing Education and Parenting: The Theory of Growth and Development has practical implications for education and parenting. By understanding the stages and patterns of development, educators and parents can tailor their approaches to accommodate the unique needs and abilities of individuals at different ages. This knowledge helps in designing appropriate learning environments, setting developmentally appropriate goals, and fostering healthy parent-child relationships.

3. Guiding Intervention and Support: The Theory of Growth

and Development informs intervention strategies for individuals with developmental delays, learning disabilities, or psychological challenges. It provides a framework for identifying areas of difficulty, understanding the underlying developmental processes, and designing targeted interventions to support growth and overcome obstacles. This knowledge aids in promoting optimal development and enhancing individuals' quality of life.

4. Shaping Social and Cultural Understanding: The Theory of Growth and Development contributes to our understanding of how individuals are shaped by their social and cultural contexts. It emphasizes the influence of societal and environmental factors on development and highlights the importance of social interactions, relationships, and cultural values in shaping an individual's growth. This knowledge promotes cultural sensitivity, fosters empathy, and encourages the creation of inclusive and supportive environments.

Conclusion: The Theory of Growth and Development illuminates the progressive and sequential changes that occur in living organisms throughout their lifespan. It provides insights into the physical, cognitive, social, and emotional aspects of development, shaping our understanding of the remarkable journey of life. The theory is vital for understanding human development, informing education and parenting practices, guiding interventions, and fostering social and cultural understanding. The Theory of Growth and Development serves as a guiding framework, helping us appreciate the complexity and diversity of human growth and maturation, and enabling us to support individuals on their unique developmental paths.

PART 5

GEOLOGY

GEOLOGY

Geology is the scientific study of the Earth, including its materials, the processes that act upon them, the structures formed by these processes, and the history of the planet and its life forms. It explores various aspects like the formation of mountains, occurrence of earthquakes, eruption of volcanoes, and the evolution of Earth's climate. Specialized branches of geology investigate specific domains, such as paleontology (study of fossils), mineralogy (study of minerals), and seismology (study of earthquakes), among others.

The importance of geology is diverse. First, it plays a crucial role in locating and managing Earth's natural resources, including fossil fuels, minerals, and groundwater. This is vital for industries such as energy production, construction, and manufacturing. Second, understanding geological processes and Earth's history is essential for predicting and mitigating natural disasters like earthquakes, volcanic eruptions, and landslides. It's also instrumental in tackling environmental issues like climate change and soil erosion. Third, geology is key to exploring other planets and celestial bodies, contributing to our broader understanding of the universe. Lastly, by tracing the evolution of life and environmental conditions on Earth, geology offers critical insights into biodiversity and the sustainability of ecosystems. In essence, the study of geology helps us responsibly and sustainably interact with our planet, providing resources for our civilization and protection from environmental hazards.

32

THEORY OF PLATE TECTONICS

The Theory of Plate Tectonics is a fundamental concept in geology that explains the movement and interactions of Earth's lithospheric plates, shaping the planet's surface and influencing geological processes. It proposes that Earth's lithosphere, Earth's outer most layer, is divided into several large and small plates that move relative to each other

Definition of the Theory of Plate Tectonics: The Theory of Plate Tectonics posits that Earth's lithosphere is divided into rigid plates that float and move on the underlying semi-fluid asthenosphere. These plates are in constant motion, driven by the forces of convection within the Earth's mantle. Plate boundaries are areas where plates interact, leading to various geological phenomena such as earthquakes, volcanic activity, and the formation of mountain ranges.

Imagine the Earth's outer surface as many large plates that make up this layer. The Theory of Plate Tectonics tells us that these plates float on the semi-fluid layer beneath them, known as the asthenosphere. Over millions of years, these plates move around, sometimes colliding, pulling apart, or sliding against each other.

These movements are like a slow dance of the plates, and they have a massive impact on the Earth's surface. They cause earthquakes, volcanic eruptions, the formation of mountains, and the shifting of continents over geological time.

For example, imagine two plates colliding, like when India slammed into Asia millions of years ago. This collision created the Himalayas, the highest mountain range on Earth.

The Theory of Plate Tectonics is crucial because it helps us understand the dynamic nature of our planet and the forces that shape its surface. It explains why earthquakes and volcanoes occur in specific regions and why continents are not fixed in place but have drifted over time.

By studying plate tectonics, scientists gain insights into Earth's geological history and the processes that continue to shape our planet. It is a fundamental concept that has revolutionized our understanding of geology and the way we perceive the dynamic nature of the Earth.

Credit to Alfred Wegener and the Contributions of Others: The Theory of Plate Tectonics builds upon the pioneering work of Alfred Wegener, a German meteorologist and geophysicist. In the early 20th century, Wegener proposed the theory of continental drift, suggesting that continents had once been connected in a single supercontinent called Pangaea. However, it was not until the mid-20th century that advancements in geophysics and the accumulation of evidence from various scientific fields led to the development of the comprehensive Theory of Plate Tectonics as we know it today.

Example of the Theory of Plate Tectonics in Real Life: An example of the Theory of Plate Tectonics in real life is the formation of mountains. The collision between the Indian and Eurasian plates is responsible for the formation of the Himalaya's, a towering mountain range. The convergence of these two plates resulted in the uplift of vast regions, causing the Earth's crust to crumple and fold forcing the now deformed plates upward, ultimately creating the majestic peaks of the Himalayas. This example demonstrates how the Theory of Plate Tectonics explains the formation of mountain ranges through plate interactions and highlights the dynamic nature of Earth's geology.

Importance of the Theory of Plate Tectonics: The Theory of Plate Tectonics holds significant importance for several reasons:

1. Understanding Earth's Geology: The Theory of Plate Tectonics provides a comprehensive framework for understanding Earth's geological processes. It explains the formation of mountains, the occurrence of earthquakes and volcanic activity, the creation of ocean basins, and the distribution of geological features. By studying plate boundaries and the movement of plates, scientists gain insights into the dynamic nature of Earth and its ever-changing surface.

2. Explaining Geological Hazards: The Theory of Plate Tectonics helps in understanding and predicting geological hazards such as earthquakes, volcanic eruptions, and tsunamis. Plate boundaries, where the interaction between plates occurs, are often associated with these hazards. By studying plate tectonics, scientists can identify areas of higher risk and develop strategies for disaster preparedness, mitigation, and response, ultimately saving lives and minimizing damage.

3. Unveiling Earth's History: The Theory of Plate Tectonics provides clues to Earth's past. By reconstructing the move-

ments of plates over millions of years, scientists can unravel the geological history of our planet. It helps explain the formation and breakup of supercontinents, the opening and closing of ocean basins, and the evolution of landforms and geological features. This knowledge aids in piecing together the puzzle of Earth's geological past.

4. Informing Resource Exploration: The Theory of Plate Tectonics guides resource exploration efforts. Plate boundaries often host mineral deposits and energy resources such as oil and gas. By understanding plate tectonics, scientists can identify areas with high resource potential, contributing to the sustainable utilization of Earth's resources.

Conclusion: The Theory of Plate Tectonics is a cornerstone of modern geology. It explains the movement and interactions of Earth's lithospheric plates, providing insights into the formation of mountain ranges, earthquakes, volcanic activity, and the evolution of Earth's surface. The theory enables us to understand Earth's dynamic nature, predict geological hazards, unveil its geological history, and inform resource exploration efforts. The Theory of Plate Tectonics has revolutionized our understanding of the Earth, shedding light on the fascinating interplay of forces that shape our planet's landscape and drive geological processes.

33

THEORY OF THE EVOLUTION OF THE EARTH

The Theory of the Evolution of the Earth explores the long and intricate journey of our planet, encompassing its geological, climatic, and biological changes over billions of years. It provides a framework to understand the processes that have shaped Earth's surface, atmosphere, and the emergence of life.

Definition of the Theory of the Evolution of the Earth: The Theory of the Evolution of the Earth describes the long-term changes that have occurred on our planet over its history. It encompasses the geological, climatic, and biological transformations that have shaped Earth's surface, atmosphere, and biosphere. This theory encompasses a range of scientific disciplines, including geology, paleontology, climatology, and evolutionary biology.

Imagine the Earth as a grand storyteller, narrating its own epic tale of change and adaptation. According to this theory, around 4.6 billion years ago, our planet was a molten mass, a fiery ball of swirling elements and gases. Over time, it began to cool and solidify, forming a rocky surface and an atmosphere.

As the Earth's surface settled, it experienced tremendous geological processes like volcanic eruptions, tectonic movements, and the shaping of mountains and valleys. The continents slowly drifted and collided, creating new landforms and changing the face of our world.

Throughout this incredible journey, life began to emerge in the form of simple organisms, such as microbes and tiny plants. These early life forms adapted and evolved in response to the changing environment, leading to the rich diversity of life we have today.

Over millions of years, more complex life forms, including dinosaurs and mammals, appeared and disappeared, leaving their mark on the Earth's history.

The Theory of the Evolution of the Earth helps us understand how the Earth itself has evolved and evolved life throughout its long history. It's a captivating story of geological forces and the rise and fall of life, showing us the beauty and resilience of our ever-changing planet.

Credit to James Hutton, Charles Lyell, and Modern Contributions: James Hutton, a Scottish geologist in the 18th century, laid the groundwork for the understanding of geological processes and the concept of deep time, contributing to the theory's foundation. Charles Lyell, a British geologist in the 19th century, expanded on Hutton's work and emphasized the principles of uniformitarianism, which propose that the same natural laws and processes observable today have shaped Earth's history. Modern contributions from scientists across various disciplines, such as geology, paleontology, and climatology, have further advanced our understanding of the Earth's evolution.

Example of the Theory of the Evolution of the Earth in Real Life: An example of the Theory of the Evolution of the Earth in real life is the study of fossil records and sedimentary layers. By examining fossils found in different geological strata, scientists can trace the progression and diversification of life on Earth over millions of years. For instance, the discovery of transitional fossils, such as the Archaeopteryx, provides evidence for the evolution of birds from reptilian ancestors. This example illustrates how the Theory of the Evolution of the Earth, coupled with paleontological evidence, contributes to our understanding of the biological transformations that have occurred throughout Earth's history.

Importance of the Theory of the Evolution of the Earth: The Theory of the Evolution of the Earth holds immense importance for several reasons:

1. Understanding Earth's History: The theory provides insights into Earth's long and intricate history. It allows scientists to reconstruct the processes that have shaped the planet, including the formation of continents, the opening and closing of ocean basins, the formation of mountain ranges, and the changes in climate and atmospheric composition. By studying the Earth's evolution, scientists gain a deeper understanding of the planet's dynamic nature.

2. Tracing the Emergence and Diversification of Life: The Theory of the Evolution of the Earth is vital for understanding the emergence and diversification of life. It helps unravel the origins of different species, the mechanisms of evolutionary change, and the complex interactions between organisms and their environment. By studying the Earth's

evolution, scientists can better appreciate the biological diversity that exists today.

3. Predicting and Mitigating Environmental Changes: The theory assists in predicting and mitigating the impacts of environmental changes on Earth's ecosystems. By understanding how past climatic and geological changes have affected life, scientists can gain insights into potential future scenarios. This knowledge helps in developing strategies for conservation, mitigating the effects of climate change, and preserving Earth's biodiversity.

4. Expanding Knowledge and Inspiring Curiosity: The Theory of the Evolution of the Earth is a testament to human curiosity and our drive to understand the world around us. It continues to inspire scientific inquiry, pushing the boundaries of knowledge and opening new avenues of research. The theory fosters a sense of wonder and appreciation for the immense geological and biological history of our planet.

Conclusion: The Theory of the Evolution of the Earth provides a comprehensive framework for understanding the dynamic history of our planet. It encompasses geological, climatic, and biological transformations, enabling scientists to trace Earth's evolution over billions of years. The theory continues to be refined by modern scientific advancements. By studying the Earth's evolution, we gain insights into our planet's geological and biological diversity, its climate dynamics, and the mechanisms of evolutionary change. The Theory of the Evolution of the Earth serves as a gateway to unraveling the remarkable story of our planet and deepening our understanding of the intricate interplay of geological and biological processes that have shaped the Earth we inhabit today.

34

Theory of the Formation of Rocks

The Theory of the Formation of Rocks explores the process-es and mechanisms behind the creation, transformation, and classification of Earth's diverse rock formations. It provides insights into the origin and evolution of rocks, shedding light on the geological forces that have shaped our planet's surface over millions of years.

Definition of the Theory of the Formation of Rocks: The Theory of the Formation of Rocks describes the scientific understanding of how rocks are created, transformed, and classified. It encompasses various processes such as sedimentation, lithification, metamorphism, and igneous activity. This theory encompasses the study of rock types, their composition, and their relationship to geological environments and processes.

Imagine rocks as ancient sculptures, slowly shaped and molded by nature's artistic hands. According to this theory, rocks can form in different ways. One common process is called the "rock cycle." It's like a never-ending story, where rocks continuously change from one type to another.

It starts with magma, which is hot, molten rock beneath the Earth's surface. When magma cools and solidifies, it becomes igneous rock. This process can happen deep underground or on the Earth's surface during volcanic eruptions.

Next, the igneous rock can undergo a transformation. It might be exposed to weathering and erosion, like gentle brushing on a sculpture, which breaks it down into smaller pieces called sediment. These sediments can be carried away by wind, water, or ice.

As the sediments settle and pile up, they get pressed and compacted by the weight above them, turning into sedimentary rock. It's like making layers upon layers of rock, just like adding paint to a canvas.

The story doesn't end there. Over time, heat and pressure can change igneous or sedimentary rocks into metamorphic rocks. It's like a metamorphosis, where rocks undergo a transformation in response to the changing conditions deep within the Earth.

This cycle of rock formation and transformation continues over millions of years, shaping the beautiful landscapes we see today.

The Theory of the Formation of Rocks helps us understand how the Earth's surface has been sculpted over time by these geological processes. It's a captivating journey through the making and remaking of rocks, showcasing the dynamic nature of our planet's crust.

Credit to James Hutton, Charles Lyell, and Modern Contributions: James Hutton, a Scottish geologist in the 18th century, made significant contributions to our understanding of rock formation and the concept of deep time. His work laid the foundation for the theory by emphasizing the importance of observing geological processes and studying rock formations to unravel Earth's history. Charles Lyell, a British geologist in the 19th century, further expanded upon Hutton's principles and popularized the concept of uniformitarianism in geology. Modern contributions from scientists across various disciplines, such as geology, petrology, and geochemistry, have further enhanced our understanding of the formation of rocks.

Example of the Theory of the Formation of Rocks in Real Life: An example of the Theory of the Formation of Rocks in real life is the formation of sedimentary rocks. Sedimentary rocks are formed through the accumulation, compaction, and cementation of sediments over time. For instance, the deposition of layers of sand, silt, and clay in riverbeds or ocean basins can eventually lead to the formation of sandstone, shale, or mudstone. This example illustrates how the Theory of the Formation of Rocks explains the process by which sedimentary rocks are created through the accumulation and lithification of sediments.

Importance of the Theory of the Formation of Rocks: The Theory of the Formation of Rocks holds significant importance for several reasons:

1. Understanding Earth's Geological History: The theory provides insights into Earth's geological history by unraveling the processes and conditions that have shaped its surface over millions of years. By studying rock formations, scientists can infer past environments, climate patterns, and geological events. This knowledge helps in reconstructing Earth's history and understanding the interplay of geological processes over time.

2. Identifying Geological Resources: The Theory of the Formation of Rocks guides the identification and extraction of geological resources. Different rock types are associated with valuable resources such as minerals, fossil fuels, and building materials. Understanding the formation and distribution of rocks aids in locating and exploiting these resources sustainably.

3. Informing Environmental Studies: The theory contributes to environmental studies by providing insights into the impact of human activities on Earth's geological systems. It helps in assessing the potential for geological hazards, such as landslides or sinkholes, and understanding the long-term effects of activities such as mining or construction on the stability of rock formations and landscapes.

4. Enhancing Earth Science Education: The Theory of the Formation of Rocks forms a core component of Earth science education. It provides a foundation for understanding Earth's dynamic processes, rock types, and geological events. By studying the formation of rocks, students gain a deeper appreciation for the incredible diversity and complexity of Earth's geological heritage.

Conclusion: The Theory of the Formation of Rocks offers insights into the processes and mechanisms behind the creation, transformation, and classification of Earth's rock formations. By studying the formation of rocks, we gain a deeper understanding of Earth's geological history, the interplay of geological processes, and the origin and evolution of different rock types. The Theory of the Formation of Rocks is invaluable in reconstructing Earth's past, identifying geological resources, informing environmental studies, and enhancing Earth science education. It unravels the geological masterpieces that make up our planet's surface, inspiring awe and curiosity about the intricate geological processes that have shaped our world.

35

THEORY OF THE FORMATION OF MINERALS

The Theory of the Formation of Minerals delves into the processes and conditions that give rise to the vast array of minerals found on Earth. It explores the chemical and physical transformations that occur within the Earth's crust, shaping the composition, structure, and properties of minerals.

Definition of the Theory of the Formation of Minerals: The Theory of the Formation of Minerals explains the scientific understanding of how minerals are created, transformed, and distributed within the Earth's crust. It encompasses processes such as crystallization from magma, precipitation from solution, metamorphism, and mineral alteration. This theory explores the diverse factors that influence mineral formation, including temperature, pressure, chemical composition, and geological environments.

According to this theory, minerals are solid substances with a specific chemical composition and a unique arrangement of atoms. One common way minerals form is through the cooling and crystallization of molten rock, called magma. When magma cools slowly, tiny crystals start to grow, just like how sugar crystals form when you make rock candy.

Another way minerals can form is through the evaporation of water. In places where water evaporates, like in hot and arid environments, minerals can be left behind, like the colorful salt crystals found in dry lake beds.

Minerals can also form when hot water circulates through rocks and dissolves certain elements. As the water cools, these elements come together to create new mineral crystals, like the beautiful quartz crystals found in caves.

Additionally, some minerals can form from the remains of once-living organisms, like the shells of marine creatures transforming into limestone over time.

The Theory of the Formation of Minerals helps us understand how the Earth's dynamic processes create these valuable and fascinating substances. It's like exploring the hidden alchemy of nature, where the right ingredients and conditions lead to the birth of precious minerals, each with its own unique beauty and properties.

Credit to Friedrich Mohs, Georgius Agricola, and Modern Contributions: The Theory of the Formation of Minerals owes its foundations to early naturalists and mineralogists who laid the groundwork for understanding minerals. Friedrich Mohs, a German mineralogist in the 19th century, introduced the Mohs scale of mineral hardness, providing a means of classifying minerals based on their relative hardness. Georgius Agricola, a German scholar from the 16th century, made significant contributions to the systematic study of minerals and mining. Modern contributions from mineralogists, petrologists, geochemists, and other scientists have further advanced our understanding of mineral formation processes.

Example of the Theory of the Formation of Minerals in Real Life: An example of the Theory of the Formation of Minerals in real life is the formation of quartz crystals in hydrothermal veins. Quartz, a common mineral, can form under specific conditions of high temperature and pressure. In hydrothermal systems, hot aqueous fluids rich in dissolved silica migrate through fractures in rocks. As the fluids cool, silica molecules precipitate and arrange themselves into ordered crystal structures, resulting in the formation of quartz veins. This example illustrates how the Theory of the Formation of Minerals helps explain the specific conditions and processes that lead to the formation of minerals like quartz.

Importance of the Theory of the Formation of Minerals: The Theory of the Formation of Minerals holds significant importance for several reasons:

1. Understanding Earth's Geological Processes: The theory provides insights into the geological processes that shape the Earth's crust and influence mineral formation. By studying the formation of minerals, scientists gain a deeper understanding of the Earth's internal dynamics, such as the cooling and solidification of magma, the movement of hydrothermal fluids, and the effects of pressure and temperature on mineral stability.

2. Identifying and Characterizing Mineral Resources: The Theory of the Formation of Minerals guides the identification and characterization of mineral resources. Understanding the conditions and processes that lead to the formation of economically valuable minerals aids in prospecting, mining, and resource management. It provides essential knowl-

edge for locating mineral deposits, assessing their quality and quantity, and optimizing extraction techniques.

3. Unveiling Earth's Geological History: Minerals serve as geological records, preserving evidence of Earth's past environments, climate, and tectonic processes. The theory allows scientists to interpret the mineralogical composition of rocks and minerals, reconstructing the Earth's history and unraveling the processes that have shaped its surface and interior.

4. Contributing to Scientific Research and Industry: The Theory of the Formation of Minerals underpins scientific research in fields such as mineralogy, petrology, and economic geology. It enables scientists to explore the relationship between mineral formation and geological processes, contributing to advancements in our understanding of Earth's materials and resources. Furthermore, the theory plays a vital role in industrial applications, including the development of materials, gemstone exploration, and the production of construction materials.

Conclusion: The Theory of the Formation of Minerals provides a comprehensive framework for understanding the processes and conditions that give rise to Earth's diverse array of minerals. By studying mineral formation, scientists gain insights into Earth's geological processes, identify and characterize mineral resources, unveil the planet's geological history, and contribute to scientific research and industrial applications. The Theory of the Formation of Minerals uncovers the treasures hidden within the Earth's crust, enabling us to appreciate the remarkable variety and beauty of Earth's geological wealth.

About the Author

Matthew S. Deal is a geography scholar and author of 'The Science Handbook'. He's a devoted husband of a talented piano pedagogue and the father of a vibrant teenager, additionally, he's been a musician, a sculptor, an artist, a lifelong learner, and cancer survivor. He writes best on vacation otherwise he becomes distracted by beautifully de-signed objects such as bikes, guitars and 1980s Toyota pickups.

AFTERWORD

If you have this book in your hands, thank you for picking it up. My wish is that there is something among the pages that give you insight or inspiration. Now, go write your own book.